THE UNCONSCIOUS

The MIT Press's publishing mission benefits from the generosity of our donors, including Bob Pozen.

THE UNCONSCIOUS

A CULTURAL HISTORY FROM HIPPOCRATES TO PHILIP K. DICK AND BEYOND

ANTONIO MELECHI

THE MIT PRESS CAMBRIDGE, MASSACHUSETTS LONDON, ENGLAND

The MIT Press
Massachusetts Institute of Technology
77 Massachusetts Avenue
Cambridge, MA 02139
mitpress.mit.edu

The MIT Press would like to thank the anonymous peer reviewers who provided comments on drafts of this book. The generous work of academic experts is essential for establishing the authority and quality of our publications. We acknowledge with gratitude the contributions of these otherwise uncredited readers.

This book was set in ITC Stone and Avenir by New Best-set Typesetters Ltd. Printed and bound in the United States of America.

Library of Congress Cataloging-in-Publication Data

Names: Melechi, Antonio, author
Title: The Unconscious : a cultural history from Hippocrates to Philip K. Dick and beyond/ Antonio Melechi.
Description: Cambridge, Massachusetts : The MIT Press, [2026] | Includes bibliographical references and index.
Identifiers: LCCN 2025015925 (print) | LCCN 2025015926 (ebook) | ISBN 9780262051026 hardcover | ISBN 9780262051040 pdf | ISBN 9780262051033 epub
Subjects: LCSH: Subconsciousness
Classification: LCC BF315 .M45 2026 (print) | LCC BF315 (ebook)
LC record available at https://lccn.loc.gov/2025015925
LC ebook record available at https://lccn.loc.gov/2025015926

10 9 8 7 6 5 4 3 2 1

EU Authorised Representative: Easy Access System Europe, Mustamäe tee 50, 10621 Tallinn, Estonia | Email: gpsr.requests@easproject.com

For Julie

CONTENTS

PROLOGUE: THE MIRROR AND THE BLINDFOLD

The history of the unconscious mind winds its way through a myriad of disciplines and social spaces, and a hubbub of expert and lay voices. Beyond philosophy, psychoanalysis, and the human sciences, this wider story of the hidden self encompasses self-experiments, letters, fiction, New Age spirituality, urban legend, courtroom testimony, and commercial hype.

Inside the conceptual rubric of "the unconscious" we find a babel of confusion and disputation. To some, the unconscious is the mind's secret enabler—the ever-spinning flywheel upon which thoughts and actions become the grooves of second nature; for others, it is the black-boxed repository of our thoughts and perceptions; or the invisible author of apparently conscious choices; or a safehouse for contraband wishes and traumatic memories; or a wellspring of creative breakthrough and mental breakdown; or the hardwired circuitry of human adaptation and evolutionary schooling. And while some contemporary psychologists deny the very existence of unconscious thought, the self-help and pop psychology industries continue to print countless promissory notes vowing to unlock its sundry powers.

This multiple and contested history has given birth to a catalogue of competing terms. *Unconscious* or *preconscious*? *Subconscious* or *nonconscious*? *Subliminal* or *dissociated*? *Automatic* or *implicit*? While this expanded lexicon is partly a consequence of the specialization and boundary work that intensified in the last decades of the nineteenth century (the period during which neurology peeled away from psychiatry, and in which psychology disclaimed its philosophical forebears), it also reflects the sheer diversity of mental phenomena that provide some

kind of opening into unconscious mental activity, or intimations of what
recent investigators have seized upon as evidence of adaptive mechanisms for behavioral guidance and "thinking without thinking." As the
psychologist and philosopher William James acknowledged in *The Variety
of Religious Experience*, any list "of everything that is latent or unobserved"
might reasonably include:

all our momentarily inactive memories, and . . . all our obscurely motivated passions, impulses, likes, dislikes, and prejudices. Our intuitions, hypotheses, fancies, superstitions, persuasions, convictions, and in general all our non-rational
operations . . . our dreams . . . whatever mystical experiences we may have, and
our automatisms, sensory or motor; our life in hypnotic and 'hypnoid' conditions . . . our delusions, fixed ideas, and hysterical accidents.[1]

That such diverse phenomena have come to find a home in the *wun-
derkammer* of the unconscious remains a moot point for many schools of
psychology. It is often argued that the term, when employed as a noun
rather than adjective, serves only to diminish its currency, to sustain the
myth that the mind is home to an autonomous secondary intelligence or
hidden personality. These misgivings are understandable, but the historian of ideas finds interest in this waywardness. There is no escaping the
fact that the unconscious is a mobile, fuzzy, and notoriously imprecise
catch-all. Repeatedly pressed into the service of tenets and credos that
bear the hallmarks of pseudoscience, it has too often served as an overblown commercial strapline, or claimed unwarranted explanatory reach
into human thought and action. No single theory of the unconscious—
Freudian, cognitive, evolutionary, behavioral, or neurological—is free
from experimental artifacts or leaps of inductive reasoning.

The story of the unconscious, as traced by this anthology, reflects its
kaleidoscopic and shifting nature, giving space to philosophers, theorists
and commentators whose insights take us far beyond the grand Freudian
drama of psychic repression and self-estrangement. This larger tapestry
stretches back to ancient dream theory, taking in medieval reflections on
somnambulism and fugue states, modern clinical studies of trance and
automatisms, and ongoing debates on the nature of instinct and freewill.
Conceptually, the unconscious reveals itself as a longstanding lightning
rod for debates and controversies in psychology and philosophy. The
questions that it has sparked remain central to our understanding of the

tenuous threads that connect us to thoughts and feelings, our motives and memories. To ask questions about the unconscious is to hold a mirror to the larger and most obscure aspects of our mental lives.

*

This anthology begins with the most quotidian portal into our unconscious lives: sleep and dreams. From Mesopotamia to Egypt, Greece to Rome, the cradles of early civilization all constructed practical religions that sought guidance and healing through dreaming, and the Western investigation of the unconscious *avant la lettre* largely begins as a counterpoint to this supernatural investment in the dream as prophecy. As far back as the writings of Aristotle and Democritus we find the first serious deliberations on the physical origins and personal significance of the dream state.

Ignoring the fact that as much as a third of our lives is spent asleep—commuting between the vivid mentation of REM sleep and quiescent phases of non-REM sleep—there might be another reason for giving chronological priority to the mysteries of the mind at rest. Surveying the earliest archaeological evidence of human tool and mark making, evolutionary psychologists have indicated that the growth of language and technology occurred in tandem with changing patterns of sleep and dreaming. Between 70,000 and 40,000 BC, our early human ancestors transitioned from polyphasic sleep (multiple rest-activity cycles over the course of a 24-hour period) to monophasic sleep (resting at night, remaining awake by day). The adjustment was slow and uneven, and its primary effect was the temporal alignment of social groups. But the uplift in REM dreaming also plausibly helped to boost learning power, consolidate memories, build social intelligence, and stimulate mental prospection. Dreams may or may not be intrinsically meaningless; the act of dreaming is definitely not.

While the Christianized and Platonic discourse of the *soul* remained predominant throughout Europe for almost 2,000 years after Aristotle, psychological thinking about automaticity, mental alterity, and divided consciousness bubbled up in unexpected quarters. As early as the fourth century, Saint Augustine asked, in his *Confessions*, whether he should be properly considered the author of his dreaming thoughts. In the same vein, Augustine, Catholicism's first introspective psychologist, went on to astutely observe his inability to understand all that he really was, marveling at the conflicting instincts and impulses that sabotaged his best intentions.

Medieval scholars were not oblivious to such unwilled operations of thought and habit. In addition to deliberating upon the power and reach of intuition and fantasy, philosophers and theologians continued to examine the fugitive aspects of consciousness in the context of delusions, visions, seizures, rapture, and somnambulism. Still, the notion of the unconscious as a curtained, intrapsychic engine-room was destined to gain sustained intellectual traction only in the last decades of the eighteenth century, as various branches of Enlightenment philosophy wrestled head-on with the problem of consciousness itself.

Cartesian dualism threw a wrench into this nascent secular psychology of mind. Contriving to regard the mind-body as a kind of supernatural automaton, a robot with a God-given soul, René Descartes envisioned the mind as an all-seeing mirror, its polished surface capturing every mental scintillation: "All my thoughts are evident to me," Descartes flatly declared in his *Meditations*, apparently forgetting the series of puzzling dreams that he had struggled to make sense of some twenty years earlier.[2] For a true psychology of the unconscious to emerge, the blind spots in Descartes' mind-mirror needed proper attention.

While an understanding of the mind's ability to perceive without awareness can be found at least as far back as Plotinus, the third-century Greek founder of Neoplatonism, the modern inquest began more than a thousand years later, some fifty years after Descartes' death. At its forefront was the philosopher and mathematician Gottfried Wilhelm Leibniz, co-inventor of calculus, who probed the "insensible perceptions" which were beyond the reach of our senses, yet central to our knowledge of the world around us:

Besides, there are hundreds of indications leading us to conclude that at every moment there is in us an infinity of perceptions, unaccompanied by awareness or reflection; that is, of alterations in the soul itself, of which we are unaware because these impressions are either too minute and too numerous, or else too unvarying, so that they are not sufficiently distinctive on their own.[3]

German intellectuals led the slow march into the *terra incognita* of the unconscious (the term *das Unbewusste* was coined by the philosopher and physician Ernst Platner in 1776). Building upon Leibniz's insights into mental phenomena without awareness, a procession of thinkers, including Kant, Hegel, Spinoza, Fichte, Schopenhauer, and Nietzsche, all expanded the psychological remit of the newly divided self, though not always in a register that helped bring clarity to their speculations. For opacity and obscurity, Kant was in a class of his own. As Schopenhauer observed, Kant's attempts to explain how intuition regulated most of our conscious thinking were most notable for the fact that no one had yet been able to make sense of them.

The most incisive of all these commentators was Friedrich Nietzsche, mustachioed firebrand and card-carrying nihilist. Proposing that consciousness was "a more or less fantastic commentary on an unknown, perhaps unknowable, but felt text," Nietzsche's aphoristic philosophy tapped, hammered, and pried open the door of the unconscious, anticipating many of the insights that Freud later channeled into the repressive foundations of psychoanalysis. One of the most famous of Nietzsche's laconic insights described how our memories are ready to function as agents of self-deception: "'I did that,' says my memory. 'I could not have done that,' says my pride, and remains inexorable. Eventually—the memory yields."[4] Consciousness, Nietzsche concluded, was no more than "a vast and thorough corruption, falsification, superficialization, and generalization."[5]

Yet the very notion of the unconscious was also a point of retreat, an intellectual haven for those at odds with Enlightenment materialism. The *Naturphilosophie* of idealist philosopher Friedrich Schelling, for example, posited an "eternal unconscious" as the hidden force behind our spontaneous activities, "the invisible root of which intelligence itself is only an expression."[6] Schelling's brand of scientific mysticism proved hugely influential. Crossing the English Channel, it became a touchstone for Romanticism's preoccupation with the irrational, creative, and chthonic forces behind our ordinary selves. A century on, echoes of Schelling's eternal unconscious continued to rebound through the emergent evolutionism of Jesuit biologist Pierre Teilhard de Chardin, New Age philosophy, the Gaia Hypothesis, and futurist theories of the global brain.

With the advent of mesmerism, the system of trance healing developed by the Viennese physician and mystagogue Frantz Anton Mesmer, the metaphysical unconscious gained ascendancy. In 1785, the curative effects of the magnetic-like substance that Mesmer claimed to harness were rejected by two scientific commissions, but the trance state became widely regarded as an instrument for expanding human thought and vision, offering, to some, direct proof of the spirituality of the soul as well as a response to objections against its immortality. Moreover, mesmerism quickly drew red-capped revolutionaries and well-heeled aristocrats into its gnostic fold. Championed across Europe and America as a natural and democratic alternative to elite medicine, the trance state inspired a heaving bookshelf of mystical tracts and medical manifestos.

The extent of scientific and public interest in the unconscious during the mid-to-late nineteenth century cannot be overstated. The psychology of mesmeric-cum-hypnotic suggestion continued to be hotly debated, creating professional schisms and sensationalized panics over its possible criminal applications. Evolutionary thought, from Lamarckism onwards, turned a spotlight on the formation of instincts and inherited characteristics. Physiologists such as William Carpenter and Benjamin Brodie studied the currents of "unconscious cerebration," using the new language of nerve-force and cortical excitation. And while psychical research explored more questionable manifestations of the second self in and out of the seance-room, neurologists such as Charcot, Freud, and Janet turned their attention to hysteria, multiple personality, and dissociative states, finding that diseases of the mind offered keyhole glimpses of latent memories, thoughts, and personalities.

The concept of the unconscious also came to acquire a social dimension. In 1895, the year in which Sigmund Freud and Josef Breuer laid out the rudiments of the "talking cure" in *Studies in Hysteria*, the French sociologist Gustave Le Bon published *The Crowd: A Study of the Popular Mind*. "In the collective mind the intellectual aptitudes of the individuals, and in consequence their individuality, are weakened," observed the mob-fearing Le Bon, taking fright at ". . . the unconscious qualities [that] obtain the upper hand."[7] Meanwhile, fellow sociologist Emile Durkheim examined this same duality in a more sober key. Proposing the existence of two aspects to our interior life, Durkheim found that the psychic forces of modern individualism were at war with group conscience and collective allegiance. Every man and woman was—as Freud would also attest—a conflicted *homo duplex*.

Unsurprisingly, the welter of theories and phenomena associated with the unconscious also met resistance, not least within the nascent field of experimental psychology. Only a decade before underwriting the psychological reach of latent processes in *The Varieties of Religious Experience*, William James had presciently warned that the notion of the unconscious was fast becoming "the sovereign means for believing what one likes in psychology and of turning what might become a science into a tumbling ground for whimsies."[8]

Skepticism of this kind was redoubled by the adoption of the unconscious as a religious or spiritualized proxy. Following the principles laid out by Wilhelm Wundt, founder of the first psychological laboratory at Leipzig, scientific psychology attacked the "mystical" theorists of the unconscious but maintained an experimental focus on the unthinking habits and processes that were central to efficient working and learning. "Rapid thought and quick action make all the difference between success and failure," wrote the Yale psychologist Edward Scripture. "A man who can think and act in one half of the time that another man can, will accumulate mental and material capital twice as fast."[9]

Pierre Janet was one of several continental psychologists who gave short shrift to the notion that the unconscious should be regarded as a source of intellectual or moral authority. In the late 1880s, after studying under Jean-Martin Charcot, the so-called "Napoleon of Neuroses," Janet undertook experiments in automatic writing, using entranced hysterics at the psychological laboratory of Salpêtrière Hospital as his subjects. What Janet found fell far short of the claims made by trance prophets and mediums famous for their freewheeling philosophies and medical diagnoses. Though some of his patients demonstrated a degree of hypermnesia—showing heightened recall for prior fugue episodes in their secondary states—their scribblings were perfunctory and bathetic. If a secondary consciousness did exist, it was, Janet surmised, notable for its dull-witted mundanity rather than its mythopoetic fluency.

Instead of rebuking the heterodox movements that found succor in the new unconscious, William James was destined to join them. The trajectory of James' thinking in the 1890s can be partly explained through his transitioning from psychology to philosophical pragmatism, but it was also intimately connected to his first-hand experiences of mind cure, hypnotism, and psychical research. The flourishing demi-mondes of alternative medicine and religion introduced James to phenomena that led him to imagine the existence of "diffuse soul-stuff" that entered into "the weak spots in the armor of the human mind."[10] In this regard, James' psychic odyssey is instructive: it reminds us that alongside the intellectual history of the unconscious, there is often a personal and social history that unfolds in tandem.

Most recent scholarship on the twentieth-century history of the unconscious has, of course, been dominated by psychoanalysis. Hundreds of books, thousands of papers, have chronicled how Freud's baroque system of mental hydraulics became a central canon of psychology and dynamic psychiatry, before evolving into what W. H. Auden called a "whole climate of opinion." To a large extent, this focus is understandable. The Freudian empire runs far and wide, it has literary merits, seeded a new profession, cultivated an enduring lexicon of the self, and provided a starting point for divergent theories such as Jung's archetype-laden unconscious and Jacques Lacan's gnomic deliberations on the "discourse of the Other." Even so, there can be no denying that intellectual history's lasting fascination with the Freudian estate has often served to obscure, even misrepresent, the wider history of the unconscious.

Contrary to the impression conveyed by Henri Ellenberger's seminal *Discovery of the Unconscious*, Freud and his followers, clinical and otherwise, neither led nor monopolized the modern investigation of the unconscious mind. Indeed, as Freud glancingly acknowledged in his 1911 essay "The Unconscious," it should be remembered that "the repressed does not constitute the whole of the unconscious."[11] Freud's proprietorial masterstroke was to transform a minority shareholding into a controlling interest, making the psychoanalytic theory of the unconscious (which soon jettisoned its model of internal censorship for the triangulated powerplay of the id, ego, and superego) synonymous with the unconscious writ large.

The patchwork of human sciences that converged upon the larger territory of the unconscious revealed itself in the spring of 1927, when a panel of invited experts came to the Illinois Society for Mental Hygiene's symposium on "The Unconscious." Among the speakers was Charles Manning Child, a zoologist best known for his work on regeneration in flatworms; Edward Sapir, world-renowned anthropologist and linguist; Kurt Koffka, psychologist and co-founder of the Gestalt School; the Left-leaning sociologist William Thomas; clinical psychologist Frederic Wells; the psychoanalytically inclined psychiatrist William A. White; and founder of behaviorism, John B. Watson, a flamboyantly vehement anti-Freudian. If the symposium was notable for its diverging constellation of hard and soft science, so too was the ready sense of ownership underwriting this wider compass.

The 1920s ushered in a golden age for the reformulation and commercialization of the unconscious. Though the decade is from a psychological perspective best remembered for the reflex arc of behaviorism (which located its unconscious in the ill-defined dash that linked stimulus and response), this was the era in which mental testing promised to reveal subconscious aspects of personality, digging deep into underlying complexes and pathology. And while insights gleaned from the likes of the Rorschach and Thematic Apperception Test began to gain acceptance in clinics, workplaces, and courtrooms across Europe and America, state-of-the-art gadgetry began to promise the middle classes *subliminal* and *subconscious* shortcuts to health and prosperity.

Among the new consumer devices with unconscious applications was the psycho-phone, the brainchild of Alois Benjamin Saliger, a Czech-born serial inventor based in New York. Essentially a phonograph connected to a wind-up alarm clock, the psycho-phone played a dozen or so specially produced "affirmation records"—with titles such as Prosperity, Normality, and Mating—which, according to the testimonials that Saliger featured in his promotional literature, were capable of affecting profound life changes. Success in business. Newfound wellbeing. Confidence. Charisma. The possibilities afforded by the psycho-phone's nightly exhortations were, it appeared, unlimited.

Advances in molecular genetics, insights into brain function and plasticity, and the birth of computational models of the mind were among the many developments that helped drive the ongoing reconceptualization of the unconscious. Alongside these emerging fields, a more knotted and contentious outgrowth of applied know-how was also being hothoused by the US military and intelligence services' bankrolling of research on mind control and behavior modification. Cold War paranoia gave credence to the idea that an individual's memory, freewill, and belief system could all be effectively breached by interventions below the level of consciousness. Brainwashed assassins and Manchurian candidates, stalwarts of the Hollywood B-movie, were among its factitious offspring.

Of all the factors impacting upon mid-century theories of the unconscious, new technology had the most far-reaching effects. At the end of the 1920s, the Electroencephalogram (EEG) machine, invented by Hans Berger, offered a graphic picture of electrical activity within the brain's vast aggregation of cells. As clinical and experimental EEG tracings showed that brain activity was associated with four bands of activity—allowing the electrical signatures of sleep, wakefulness, epilepsy, and a range of borderline states to be recorded—the unconscious began to be *visually* transformed. With the emergence of MRI scanning technology, neuronal activity was destined to become a pixelated proxy for unconscious processes, helping to revive what Daniel Dennett calls the "Cartesian Theatre," in which the mind performs the part of an epiphenomenal ghost in the substratum of the brain.

The impact of technology and innovation upon the modern mind goes far beyond the clinic and laboratory. In the first decades of the twentieth century, the philosophers Georg Simmel and Walter Benjamin each explored the psychology of the metropolis, examining the "distracted attention" that city life fostered; the Harvard psychologist Hugo Munsterberg applied himself to the ways in which cinema's flickering illusion hotlined processes outside the spectator's awareness; and as the popular press reported the first accounts of the fugue experiences associated with long-distance car journeys, industrial psychologists were called on to find ways of mitigating the monotony of repetitive factory work. Daydreaming had become an economic issue.

Every media development that has found a place in the social sphere, from radio and television to the PC and mobile phone, has heightened and redoubled what the social psychologist John Bargh dubs "the automaticity of everyday life."[12] At the beginning of the twenty-first century, what we think and feel, do and say, is increasingly managed by digital technologies that hide in plain sight, and by algorithms that invisibly prime and prompt our on and off-screen decision-making. The claims of philosophers and psychologists who contend that the mind is best understood as an interfacing and "extended cognitive system" have become difficult to ignore. More than ever, unthinking repetition and reinforcement, the twin pillars of instrumental conditioning, are the binding agents in human relationships with media technology.

No single anthology can hope to map the maze of intellectual and social byways that the unconscious traverses. Over the following pages, I have simply tried to trace some of the major and minor destinations in its Protean journey across 25 or so centuries of Western thought. Surveying this large vista of ideas and experiments, I have had to be highly selective, and there are many notable thinkers and traditions that have, for reasons of space alone, been omitted. The whole galaxy of Eastern thought on non-conscious states of mind, spanning Hindu philosophy and the yogic schools, Buddhism, and Daoism, is only touched upon in my final chapter. The anthropology of trance and cross-cultural analogues of the unconscious are given only glancing acknowledgement. And several well-known modern commentators and practitioners from various fields have been overlooked to give voice to lesser-known figures.

In the following chapters, I have worked in synoptic outline, setting key experiments and philosophical insights alongside neglected doctrines, discarded technologies, and little-known ephemera. At all times, I treat the stories, myths, theories, and speculations that have crowded around the unconscious as equally admissible evidence, and the following compendium of extracts—leaning always towards the pithy and non-technical—works to emphasize the status of the unconscious as a social, philosophical, and scientific object. Hovering between discovery and invention, the unconscious provides us with what the literary critic Kenneth Burke describes as a terministic screen, a teeming body of "terms through which humans perceive the world, and that direct attention away from some interpretations and toward others."[13] At the forefront of today's so-called "new unconscious" stands a broad alliance of cognitive scientists, evolutionary psychologists, philosophers, and neuroscientists. Meanwhile, prospecting its commercial subsoil, there is a booming nexus of cyberneticists, behavioral engineers, right-brain coaches, neurofeedback experimenters, human potential gurus, neuro-linguistic programmers, and implicit bias trainers.

While the language and grammar of the unconscious has helped to frame the various ways in which our minds and bodies are routinely directed without awareness and intentionality, there is no escaping its ersatz detours and disciplinary misadventures. From psychoanalysis to Scientology, self-help to biofeedback adventures in "alpha power," the unconscious has served an array of questionable masters and dubious ends. Appropriately enough, its very own history turns out to be one of blindness and insight, wisdom and folly.

1

NIGHT SCHOOL: THE PHILOSOPHY OF DREAMS

"The waking have one common world," observed the fifth-century Greek philosopher Heraclitus, "but the sleep turn aside each into a world of his own."[1] Heraclitus' observations on the solipsistic nature of sleep may strike us as a psychological truism—what, after all, could be more intimate than the thoughts that visit us, seemingly unbidden, when our eyes are closed, bodies paralysed, and minds impervious to sensation?—but the history of the dreaming mind is by no means a private or subjective affair. Within the apparent outlandishness and absurdity of dreams lies a common mystery, an enigma that has captivated a procession of soothsayers, philosophers, and scientists, taking us from the magical crucible of the ancient dream temple to the electrified cradle of the modern-day sleep lab.

Dreams held a more significant place in Greek civilization than they did in the Roman world. At hundreds of temples and sacred sites across rural and urban Greece, ordinary people came to experience dreams that were considered gifts from the gods. Through these dreams, the ill and the infirm might be spontaneously returned to good health, or receive priestly prescriptions based on the nature of their spoken dreams. In either case, the ritual of dream incubation made for a deeply emotional and cathartic experience, which with good reason has been described as a form of primitive psychoanalysis. "One's hair stood on end; one cried and felt happy; one's heart swelled but not with vainglory," wrote Aristides in the second century AD, following his conversion to the cult of Asclepius.[2]

Aristotle had by this time proposed that dreams held no supernatural significance, and secular physicians such as Hippocrates and Galen followed suit, interpreting dream experiences as somatic manifestations of health, typically connected to an imbalance of the humors. Still, within popular culture, dreams were above all valued for their prophetic currency and much of what we know about dream interpretation in antiquity comes from the work of Artemidorus, a professional soothsayer who was born in the second century, in the city of Ephesus.

Artemidorus's *Oneirocritica*, the only surviving work of ancient dream interpretation, sought to defend the art of dream divination (a practice that had already become widely discredited, mainly on account of its association with its marketplace practitioners), while offering practical instruction on how to decode hundreds of allegorical elements typically found in dreams—particularly those relating to birth and death, gods, sex, slaves, flora and fauna, food and drink. Too many of his fellow soothsayers had, Artemidorus argued, grasped that only one class of dreams, *oneroi*, had supernatural import; too few appreciated that the symbolism of such dreams could be god-sent or derived from the mind's own prophetic powers. For this reason, the task of dream interpretation required knowledge of the dreamer's background, status, and profession, as well as their physical and mental health. To know the dream, to determine in what ways it might be ominous or auspicious, one must know the dreamer.

While the *Oneirocritica* was destined to remain a touchstone for almost all future works of dream interpretation, rising skepticism towards the prophetic value of the dreamwork led to much greater interest in the dream as a clouded mirror to waking thoughts. From the late medieval period onwards, scholars began to caution against "lying auguries" and "deceitful dreams," following Aristotle in linking the "somnium" to the physical and psychological state of the dreamer. Soon enough, the Swiss alchemist and philosopher Paracelsus went deeper still: "That which the dream shows," he wrote, "is the shadow of such wisdom that exists in the man, even if during his waking state he knows nothing about it."[3]

During the eighteenth and nineteenth centuries, dream theories multiplied, traversing new disciplines. As well as being trumpeted as a source of self-knowledge, the dreamwork was now heralded as a universal language; an analogue of insanity; a random by-product of brain chemistry at sleep. Yet perhaps the most revealing insights came from French investigators such as Alfred Maury and Marquis Hervey de

Saint-Denys, who independently studied the inner world of dreams via self-experimentation, assiduously recording the dreaming mind's ability to revive long-forgotten memories, to build complete scenes and vistas from simple hypnagogic images, and, most curiously of all, to fall under conscious control.

Though Freud was well aware of Maury, Saint-Denis, and other earlier dream theorists when he wrote his most famous work, *The Interpretation of Dreams*, he claimed to have found nothing of any "scientific" value in their investigations; all his predecessors had, he maintained, allowed themselves to be beguiled by the manifest content of the dream. For Freud, the dream, once unraveled through free association, would always show itself as a repressed wish in disguise, a Trojan Horse passing through the bulwark of the ego's defense.

When Freud died in 1939, his unerring insistence that dreams were the "royal road" to understanding the unconscious mind had been widely popularized. As well as becoming a cornerstone of psychoanalytic practice, the notion of the dream as a censored wish was embraced by psychologists, anthropologists, surrealists, and Hollywood directors. But a new science of sleep was about to emerge, a science that was built around the electroencephalography (EEG) machine, an instrument that allowed researchers to see behind the curtain of sleep and record the electrical underbelly of the dream state in real time.

To begin with, sleep science was ill-equipped to provide any significant clues as to the mental or physical nature of dreaming. The general brain activity measured by way of an EEG machine could objectively determine whether a person was awake, in deep sleep, or anaesthetized, but the electrical activity produced in the dreaming state appeared virtually the same as that seen in wakefulness; rapid-eye movement was the only objective sign that a sleeper was dreaming.

As the sleep lab availed itself of new technologies, Harvard Medical School psychiatrists J. Allan Hobson and Robert McCarley were able to locate a cluster of "REM on" cells in the forebrain. The activation of these cells, they proposed, was responsible for random neural activity that the mind attempted to piece together into the dream's improvised narrative. "Dreaming is no longer mysterious," Hobson blustered, echoing the chutzpah that that Freud brought to his equally overblown dream theory.[4]

The so-called activation-synthesis model of dreaming certainly provided an important corrective to the psychodynamic approach that Freud had inaugurated, but its limitations and shortcomings were also plain to see. Was the psychological hallmark of dreams always their "bizarre" and "evanescent" nature? Was dreaming restricted to non-REM sleep? If dreams did have an adaptive function, why should this be confined to preparing the dreamer for waking consciousness?

All these questions suggested that no single theory or perspective could hope to contain the vast and multiform nature of dreaming, which is perhaps why Aristotle, writing over two thousand years earlier, had declared discussion of any of the attributes that attended sleep such "an obscure and indeterminate business."[5]

*

The Greek physician Hippocrates (460–370 BC), often considered the father of Western medicine, used his patients' dreams as an aid to diagnosis and prognosis.

HIPPOCRATES, *ON REGIMEN* (400 BC)

This is the truth of the matter. Such dreams as repeat in the night a man's actions or thoughts in the day-time, representing them as occurring naturally, just as they were done or planned during the day in a normal act— these are good for a man. They signify health, because the soul abides by the purposes of the day, and is overpowered neither by surfeit nor by depletion nor by any attack from without. But when dreams are contrary to the acts of the day, and there occurs about them some struggle or triumph, a disturbance in the body is indicated, a violent struggle meaning a violent mischief, a feeble struggle a less serious mischief. As to whether the. act should be averted or not I do not decide, but I do advise treatment of the body. For a disturbance of the soul has been caused by a secretion arising from some surfeit that has occurred. Now if the contrast be violent, it is beneficial to take an emetic, to increase gradually a light diet for five days, to take in the early morning long, sharp walks, increasing them gradually, and to adapt exercises, when in training, so as to match the gradual increase of food. If the contrast be milder, omit the emetic, reduce food by a third, resuming this by a gentle, gradual increase spread over five days. Insist on vigorous walks, use voice-exercises, and the disturbance will cease.[6]

Following its 1518 Latin translation, Artemidorus's Oneirocritica *unleashed a wave of occult dream guides and manuals, shaping the development of later dream auguries and fueling centuries of fascination with dreams as symbols freighted with hidden knowledge. In this passage, Artemidorus examined one of the most common and perplexing of dream experiences.*

ARTEMIDORUS, *ONEIROCRITICA* (2ND CENTURY AD)

Of teeth, the top-teeth signify the superior and pre-eminent people in the house of the observer, but the bottom-teeth signify inferior people. For it is necessary to consider the mouth as one's house, and the teeth the people in the house, of which those on the right signify men, and those on the left signify women, unless one of these genders is scarce in one's household, for instance, if someone who is a brothel-keeper thus has an all-female house or one who is fond of the country life an all-male house. For in these cases the teeth on the right signify [the] older men and older women, and those on the left younger men and younger women. And, moreover, the teeth called the 'cutters', that is, those in the front, signify the young, and the canines the middle-aged, and the molars signify older people [which some call 'grinders']. And whichever sort of tooth one casts away, he will be deprived of this sort of person. And since the teeth in fact not only signify people but acquisitions, it is necessary to consider the molars as signifying heirlooms, and the canines things that are not worth all that much, and the incisors one's household objects. And so it is logical that teeth that have fallen out signify the loss of one's property. And, moreover, the teeth signify the affairs relating to one's life. And, of these affairs, the molars signify secrets and things not-to-be mentioned, and the canines things not manifest to many, and the incisors things that are quite plain and that are accomplished through word and voice. And so the teeth, when they fall out, hinder their corresponding affairs.[7]

Synesius of Cyrene (373–414 AD), the Greek Bishop of Ptolemais in ancient Libya, recommended the keeping of a night book.

SYNESIUS OF CYRENE, *CONCERNING DREAMS* (CA. 405 AD)

For what could be more abundant than dreams, and what more fascinating? These induce even fools to pay heed to them. It would therefore be shameful for those who have lived ten years beyond adolescence to stand in need of any other diviner, shameful that they should not have accumulated an abundant store of technical principles. It should be a wise proceeding even to publish our waking and sleeping visions and their attendant circumstances; the things to do, I say, unless the culture of the city is like to be too rustic for so novel an enterprise. We shall therefore see fit to add to what are called "day books" what we term "night books," so as to have records to remind us of the character of each of the two lives concerned; for our argument already laid it down that certain life exists in imagination, at one moment better, at another worse than the intermediate, according to the relation of the pneuma to health and disease. If in this way, therefore, we make profitable the observation by which the art is developed, and if nothing slips our memory, in other respects also the result will be a refined pastime; it will be paying oneself the compliment of a history of one's waking and sleeping moments.[8]

The erotic and sexual content of dreams was widely and openly discussed by Greek and Roman philosophers. In Book 10 of his Confessions, *Saint Augustine (354–430 AD) reflected on his own "suggestive thoughts."*

ST. AUGUSTINE, *CONFESSIONS* (397–400 AD)

[I]n my memory of which I have spoken at length, there still live images of acts which were fixed there by my sexual habit. These images attack me. While I am awake they have no force, but in sleep they not only arouse pleasure but even elicit consent, and are very like the actual act. The illusory image within the soul has such force upon my flesh that false dreams have an effect on me when asleep, which the reality could not have when I am awake. During this time of sleep surely it is not my true self, Lord my God? Yet how great a difference between myself at the time when I am asleep and myself when I return to the waking state. Where then is reason which, when wide-awake, resists such suggestive thoughts, and would remain unmoved if the actual reality were to be presented to it? Surely reason does not shut down as the eyes close. It can hardly fall asleep with the bodily senses. For if that were so, how could it come about that often in sleep we resist and, mindful of our avowed commitment and adhering to it with strict chastity, we give no assent to such seductions? Yet there is a difference so great that, when it happens otherwise than we would wish, when we wake up we return to peace in our conscience. From the wide gulf between the occurrences and our will, we discover that we did not actively do what, to our regret, has somehow been done in us.[9]

The philosopher Thomas Hobbes (1588–1679) wondered whether his contemporary, René Descartes, was right to suggest that there was no criterion for distinguishing our waking thoughts from our dreams.

THOMAS HOBBES, *LEVIATHAN* (1651)

The imaginations of them that sleep, are those we call dreams. And these also (as all other imaginations) have been before, either totally, or by parcels in the sense. And because in sense, the brain, and nerves, which are the necessary organs of sense, are so benumbed in sleep, as not easily to be moved by the action of external objects, there can happen in sleep, no imagination; and therefore no dream, but what proceeds from the agitation of the inward parts of man's body; which inward parts, for the connexion they have with the brain, and other organs, when they be distempered, do keep the same in motion; whereby the imaginations there formerly made, appear as if a man were waking; saving that the organs of sense being now benumbed, so as there is no new object, which can master and obscure them with a more vigorous impression, a dream must needs be more clear, in this silence of sense, than are our waking thoughts. And hence it cometh to pass, that it is a hard matter, and by many thought impossible to distinguish exactly between sense and dreaming. For my part, when I consider, that in dreams, I do not often, nor constantly think of the same persons, places, objects, and actions that I do waking; nor remember so long a train of coherent thoughts, dreaming, as at other times; and because waking I often observe the absurdity of dreams, but never dream of the absurdities of my waking thoughts; I am well satisfied, that being awake, I know I dream not; though when I dream, I think myself awake.[10]

Physician-philosopher Erasmus Darwin (1731–1802), grandfather of Charles Darwin, was one of the eighteenth century's finest psychological students of the dream.

ERASMUS DARWIN, *ZOONOMIA* (1803)

The rapidity of the succession of transactions in our dreams is almost inconceivable; insomuch that, when we are accidentally awakened by the jarring of a door, which is opened into our bed-chamber, we sometimes dream a whole history of thieves or fire in the very instant of awaking.

During the suspension of volition we cannot compare our other ideas with those of the parts of time in which they exist; that is, we cannot compare the imaginary scene, which is before us, with those changes of it, which precede or follow it: because this act of comparing requires recollection or voluntary exertion. Whereas in our waking hours, we are perpetually making this comparison, and by that means our waking ideas are kept confident with each other by intuitive analogy; but this companion retards the succession of them, by occasioning their repetition. Add to this, that the transactions of our dreams consist chiefly of visible ideas, and that a whole history of thieves and fire may be *beheld* in an instant of time like the figures in a picture. . . .

Two other remarkable circumstances of our dreaming ideas are their inconsistency, and the total absence of surprise. Thus we seem to be present at more extraordinary metamorphoses of animals or trees, than are to be met with in the fables of antiquity; and appear to be transported from place to place, which seas divide, as quickly as the changes of scenery are performed in a play-house; and yet are not sensible of their inconsistency, nor in the least degree affected with surprise.

We must consider this circumstance more minutely. In our waking trains of ideas, those that are inconsistent with the usual order of nature, so rarely have occurred to us, that their connexion is the slightest of all others: hence, when a consistent train of ideas is exhausted, we attend to the external stimuli, that usually surround us, rather than to any inconsistent idea, which might otherwise present itself; and if an inconsistent idea should intrude itself, we immediately compare it with the preceding one, and voluntarily reject the train it would introduce; this appears further in the Section on Reverie, in which state of the mind external stimuli are not attended to, and yet the streams of ideas are kept consistent by the efforts of volition. But as our faculty of volition is suspended, and all external stimuli are excluded in sleep, this slighter connexion of ideas takes place; and the train is said to be inconsistent; that is, dissimilar to the usual order of nature.[11]

Pre-industrial households experienced two major intervals of nocturnal sleep, a "first sleep" and "second sleep," bridged, somewhere around midnight, by an hour or so of wakefulness in which individuals meditated, chatted, or tended to household chores. With the introduction of domestic and urban lighting, the second sleep diminished, altering the nature of sleep and dreams.

ROGER EKIRCH, *AT DAY'S CLOSE: A HISTORY OF NIGHTTIME* (2005)

Beginning in the late seventeenth century, divided slumber gradually grew less common in cities and towns, first among propertied households, then, more slowly, among other social ranks due to later bedtimes and improved illumination. Heightened exposure to artificial lighting, both at home and abroad, altered circadian rhythms as old as man himself. By the mid-1800s, only people unable to afford adequate lighting, in all likelihood, still experienced segmented sleep, particularly if forced to retire at an early hour. The working class author of *The Great Unwashed*, for example, remarked in 1868 that laborers who had "to turn out early in the morning" were "already in their first sleep" at night when the streets of his town were "still in a state of comparative bustle." Altered, too, was the relative importance of nocturnal dreams. No longer did most sleepers experience an interval of wakefulness in which to ponder visions in the dead of night. With the transition to a new pattern of slumber, at once consolidated and more compressed, increasing numbers lost touch with their dreams and, as a consequence, a traditional avenue to their deepest emotions. It is no small irony that, by turning night into day, modern technology has helped to obstruct our oldest path to the human psyche. That, very likely, has been the greatest loss, to paraphrase an early poet, of having been "disannulled of our first sleep, and cheated of our dreams and fantasies."[12]

While ostensibly "losing touch" with dreams, early nineteenth-century readers were in thrall to the secrets unlocked by a stream of affordable dream books.

CHARLES MACKAY, *EXTRAORDINARY POPULAR DELUSIONS* (1841)

It is quite astonishing to see the great demand there is, both in England and France, for dreambooks, and other trash of the same kind. Two books in England enjoy an extraordinary popularity, and have run through upwards of fifty editions in as many years in London alone, besides being reprinted in Manchester, Edinburgh, Glasgow, and Dublin. One is Mother Bridget's Dream-book and Oracle of Fate; the other is the Norwood Gipsy. It is stated, on the authority of one who is curious in these matters, that there is a demand for these works, which are sold at sums varying from a penny to sixpence, chiefly to servant-girls and imperfectly educated people, all over the country, of upwards of eleven thousand annually; and that at no period during the last thirty years has the average number sold been less than this. The total number during this period would thus amount to 330,000.[13]

French sinologist Marquis Leon Hervey de Saint-Denys (1822–1892) began recording his dreams at the age of 13, developing a unique talent for what is now called "lucid dreaming."

HERVEY DE SAINT-DENYS, *DREAMS AND HOW TO GUIDE THEM* (1867)

Let us briefly recall what we have attempted to establish both in regard to the psychology of dreams in general and in regard to the practical means of evoking or dispelling while sleeping certain idea-images, of guiding the mind in its spontaneous or voluntary movements, and finally of guiding one's dreams according to one's desires.

We have said that we did not believe in the sleep of thought, that we did not consider the exercise of any faculty to be suspended by sleep, that if attention was sometimes difficult, the will weakened and judgements erroneous in the sleeping man, in contrast imagination, memory and sensibility acquired a power of enormous expansion; such that if the dream state does not allow us to maintain the exact intellectual equilibrium indispensable to the accomplishment of a work of the mind that is reasonable in every respect, then it can at least open up on the ideal world horizons that are unknown in real life.

Three essential conditions have been pointed out in order to succeed in making oneself the master of the illusions of one's sleep.

1. To have while sleeping the awareness of one's sleep, a habit acquired fairly quickly by the simple fact of keeping a diary of one's dreams;
2. To associate certain memories with the recall of certain sensory perceptions, in such a way that the return of these sensations, administered during sleep, introduces into the midst of our dreams the idea-images that we have already made a part of them;
3. These idea-images therefore contributing to form the pictures of our dreams, to employ the will (which will never be lacking when one truly knows one is dreaming) to guide their development in accordance with the application of the principle that to think of a thing is to dream it.[14]

Edward Burnett Tylor (1832–1922) found that dreams were central to the natural religion of the "lower races," being used universally to maintain contact with the dead.

EDWARD B. TYLOR, *PRIMITIVE CULTURE* (1871)

Among the Indians of North America, we hear of the dreamer's soul leaving his body and wandering in quest of things attractive to it. These things the waking man must endeavour to obtain, lest his soul be troubled, and quit the body altogether. The New Zealanders considered the dreaming soul to leave the body and return, even travelling to the region of the dead to hold converse with its friends. The Tagals of Luzon object to waking a sleeper, on account of the absence of his soul. The Karens, whose theory of the wandering soul has just been noticed, explain dreams to be what this la [soul] sees and experiences in its journeys when it has left the body asleep. They even account with much acuteness for the fact that we are apt to dream of people and places which we knew before; the leip-pya, they say, can only visit the regions where the body it belongs to has been already. Onward from the savage state, the idea of the spirit's departure in sleep may be traced into the speculative philosophy of higher nations, as in the Vedanta system, and the Kabbala. . . .

The North American Indians allowed themselves the alternative of supposing a dream to be a visit from the soul of the person or object dreamt of, or a sight seen by the rational soul, gone out for an excursion while the sensitive soul remains in the body. So the Zulu may be visited in a dream by the shade of an ancestor, the itongo, who comes to warn him of danger, or he may himself be taken by the itongo in a dream to visit his distant people, and see that they are in trouble; as for the man who is passing into the morbid condition of the professional seer, phantoms are continually coming to talk to him in his sleep, till he becomes, as the expressive native phrase is, "a house of dreams". In the lower range of culture, range of culture, it is perhaps most frequently taken for granted that a man's apparition in a dream is a visit from his disembodied spirit, which the dreamer, to use an expressive Ojibwa idiom, "sees when asleep". Such a thought comes out clearly in the Fijian opinion that a living man's spirit may leave the body, to trouble other people in their sleep; or in a recent account of an old Indian woman of British Columbia sending for the medicine man to drive away the dead people who came to her every night. A modern observer's description of the state of mind of the negroes of South Guinea in this respect is extremely characteristic and instructive. "All their dreams are construed into visits from the spirits of their deceased friends. The cautions, hints, and warnings which come to them through this source, are received with the most serious and deferential attention, and are always acted upon in their waking hours. The habit of relating their dreams, which is universal, greatly promotes the habit of dreaming itself, and hence their sleeping hours are characterized by almost as much intercourse with the dead as their waking are with the living. This is, no doubt, one of the reasons of their excessive superstitiousness. Their imaginations become so lively that they can scarcely distinguish between their dreams and their waking thoughts, between the real and the ideal, and they consequently utter falsehood without intending, and profess to see things which never existed."[15]

The idea of "unconscious cerebration" provided Victorian commentators such as Frances Power Cobbe (1822–1904) with a popular framework for the discussion of dreams.

FRANCES POWER COBBE, "DREAMS AS ILLUSTRATIONS OF INVOLUNTARY CEREBRATION" (1872)

A correspondent has kindly sent me the following interesting remarks on the above:—"When dropping asleep some nights ago I suddenly started awake with the thought on my mind, 'Why I was *making* a dream!' I had detected myself in the act of inventing a dream. Three or four impressions of scenes and events which had passed across my mind during the day were present together in my mind, and the effort was certainly being made, but not by my fully conscious will, to arrange them so as to form a continuous story. They had actually not the slightest connnexion, but a process was evidently going on in my brain by which they were being united into one scheme or plot. Had I remained asleep until the plot had been matured, I presume my waking sensation would have been that I had an ordinary dream. But perhaps through the partial failure of the unconscious effort at a plan, I woke up just in time to catch a trace of the 'unconscious cerebration' as it was vanishing before the full light of conscious life. I accordingly propounded a tentative theory to my friends, that what takes place in dreams—a sort of faint shadow of the mind's natural craving for and effort after system and unity. Your explanation of dreams by reference to the 'myth-making tendency,' seems to be so nearly in accord with mine that I venture to write on the subject."[16]

Though Friedrich Nietzsche (1844–1900) was hardly the first philosopher to consider dreams as a kind of education in primitive thinking, he carried the idea further than most.

FRIEDRICH NIETZSCHE, *HUMAN, ALL TOO HUMAN* (1878)

The function of the brain that sleep encroaches upon most is the memory: not that it ceases altogether—but it is reduced to a condition of imperfection such as in the primeval ages of mankind may have been normal by day and in waking. Confused and capricious as it is, it continually confuses one thing with another on the basis of the most fleeting similarities: but it was with the same confusion and capriciousness that the peoples composed their mythologies, and even today travellers observe how much the savage is inclined to forgetfulness, how his mind begins to reel and stumble after a brief exertion of the memory and he utters lies and nonsense out of mere enervation. But in dreams we all resemble this savage; failure to recognize correctly and erroneously supposing one thing to be the same as another is the ground of the false conclusions of which we are guilty in dreams; so that, when we clearly recall a dream, we are appalled to discover so much folly in ourselves.—The perfect clarity of all the images we see in dreams which is the precondition of our unquestioning belief in their reality again reminds us of conditions pertaining to earlier mankind, in whom hallucination was extraordinarily common and sometimes seized hold on whole communities, whole peoples at the same time. Thus: in sleep and dreams we repeat once again the curriculum of earlier mankind.[17]

William Dean Howells (1837–1920), known as "The Dean of American Letters," delights in the banality of other people's dreams.

WILLIAM DEAN HOWELLS, *TRUE, I TALK OF DREAMS* (1895)

Everyone knows how delightful the dreams are that one dreams one's self, and how insipid the dreams of others are. I had an illustration of the fact, not many evenings ago, when a company of us got telling dreams. I had by far the best dreams of any; to be quite frank, mine were the only dreams worth listening to; they were richly imaginative, delicately fantastic, exquisitely whimsical, and humorous in the last degree; and I wondered that when the rest could have listened to them they were always eager to cut in with some silly, senseless, tasteless thing that made me sorry and ashamed for them. I shall not be going too far if I say that it was on their part the grossest betrayal of vanity that I ever witnessed.[18]

The well-known child psychologist James Sully (1842–1923) developed an intriguing theory of the dream as a psychic time machine, a resurrector of "dead selves."

JAMES SULLY, "THE DREAM AS A REVELATION" (1893)

The proposition that the soundest of men undergo changes of personality may well strike the reader as paradoxical; yet the paradox is only on the surface. Although we talk of ourselves as single personalities, as continuing to be the same as we were, a little thought suffices to show that this is not absolutely true. . . .

Now our dreams are a means of conserving these successive personalities. When asleep we go back to the old ways of looking at things and of feeling about them, to impulses and activities which long ago dominated us, in a way which seems impossible in waking hours, when the later self is in the ascendant. In this way the rhythmic change from wakefulness to sleep affects a recurrent reinstatement of our 'dead selves,' an overlapping of the successive personalities, the series of whose doings and transformations constitutes our history.

There is one other way in which dreams may become an unveiling of what is customarily hidden, viz., by giving free play to individual characteristics and tendencies. It is a commonplace that our highly artificial form of social life tends greatly to restrict the sphere of individuality. Our peculiar tendencies get sadly crossed and driven back in the daily collision with our surroundings. Much that is deepest and most vital in us may in this way be repressed and atrophied. . . . Hence according to what has been said above, they are very apt to disclose themselves when sleep has stupefied the dominant personality.[19]

Freud's "royal road to the unconscious" was a treacherous intersection of four major dream processes: condensation, secondary revision, displacement, and representation.

SIGMUND FREUD, *THE INTERPRETATION OF DREAMS* (1900)

It is true that we distort dreams in attempting to reproduce them; here we find at work once more the process which we have described as the secondary (and often ill conceived) revision of the dream by the agency which carries out normal thinking. But this distortion is itself no more than a part of the revision to which the dream-thoughts are regularly subjected as a result of the dream-censorship. The other writers have at this point noticed or suspected the part of dream-distortion which operates manifestly; we are less interested, since we know that a much more far-reaching process of distortion, though a less obvious one, has already developed the dream out of the hidden dream-thoughts. The only mistake made by previous writers has been in supposing that the modification of the dream in the course of being remembered and put into words is an arbitrary one and cannot be further resolved and that it is therefore calculated to give us a misleading picture of the dream. They have underestimated the extent to which psychical events are determined. There is nothing arbitrary about them. . . .

In analysing the dreams of my patients I sometimes put this assertion to the following test, which has never failed me. If the first account given me by a patient of a dream is too hard to follow I ask him to repeat it. In doing so he rarely uses the same words. But the parts of the dream which he describes in different terms are by that fact revealed to me as the weak spot in the dream's disguise: they serve my purpose just as Hagen's was served by the embroidered mark on Siegfried's cloak.

That is the point at which the interpretation of the dream can be started. My request to the patient to repeat his account of the dream has warned him that I was proposing to take special pains in solving it; under pressure of the resistance, therefore, he hastily covers the weak spots in the dream's disguise by replacing any expressions that threaten to betray its meaning by other less revealing ones. In this way he draws my attention to the expression which he has dropped out. The trouble taken by the dreamer in preventing the solution of the dream gives me a basis for estimating the care with which its cloak has been woven.

[*But if the dream was a disguised wish, why did so many dreams relate to recent happenings in the patient's life?*]

[In] every dream some link with a recent daytime impression—often of the most insignificant sort—is to be detected. We have not hitherto been able to explain the necessity for this addition to the mixture that constitutes a dream. And it is only possible to do so if we bear firmly in mind the part played by the unconscious wish and then seek for information from the psychology of the neuroses. We learn from the latter that an unconscious idea is as such quite incapable of entering the preconscious and that it can only exercise any effect there by establishing a connection with an idea which already belongs to the preconscious, by transferring its intensity on to it and by getting itself 'covered' by it. Here we have the fact of 'transference,' which provides an explanation of so many striking phenomena in the mental life of neurotics. . . . I hope I may be forgiven for drawing analogies from everyday life, but I am tempted to say that the position of a repressed idea resembles that of an American dentist in this country: he is not allowed to set up in practice unless he can make use of a legally qualified medical practitioner to serve as a stalking-horse and to act as a 'cover' in the eyes of the law. And just as it is not exactly the physicians with the largest practices who form alliances of this kind with dentists, so in the same way preconscious or conscious ideas which have already attracted a sufficient amount of the attention that is operating in the preconscious will not be the ones to be chosen to act as covers for a repressed idea. The unconscious prefers to weave its connections round preconscious impressions and ideas which are either indifferent and have thus had no attention paid to them, or have been rejected and have thus had attention promptly withdrawn from them. . . . [T]he reason why these recent and indifferent elements so frequently find their way into dreams as substitutes for the most ancient of all the dream-thoughts is that they have least to fear from the censorship imposed by resistance.[20]

*Physician and writer Havelock Ellis (1859–1939), though sympathetic to
Freud and psychoanalysis, better understood one element of dreaming that
Freud routinely neglected—its joy.*

HAVELOCK ELLIS, *THE WORLD OF DREAMS* (1911)

Dreaming is thus one of our roads into the infinite. And it is interesting
to observe how we attain it—by limitation. The circle of our conscious life
is narrowed during sleep; it is even by a process of psychic dissociation
broken up into fragments. From that narrowed and broken-up consciousness the outlook becomes vaster and more mysterious, full of strange and
unsuspected fascination, and the possibilities of new experiences, just as
a philosophic mite inhabiting a universe consisting of a Stilton cheese
would probably be compelled to regard everything outside the cheese
as belonging to the realm of the Infinite. In reality, if we think of it, all
our visions of the infinite are similarly conditioned. It is only by empha-
sising our finiteness that we ever become conscious of the infinite. The
infinite can only be that which stretches far beyond the boundaries of
our own personality. It is the charm of dreams that they introduce us
into a new infinity. Time and space are annihilated, gravity is suspended,
and we are joyfully borne up in the air, as it were in the arms of angels;
we are brought into a deeper communion with Nature, and in dreams a
man listens to the arguments of his dog with as little surprise as Balaam
heard the reproaches of his ass. The unexpected limitations of our dream
world, the exclusion of so many elements which are present even unconsciously in waking life, impart a splendid freedom and ease to the intellectual operations of the sleeping mind, and an extravagant romance, a
poignant tragedy, to our emotions. 'He has never known happiness,' said
Lamb, speaking out of his own experience, 'who has never been mad.'
And there are many who taste in dreams a happiness they never know
when awake. In the waking moments of our complex civilised life we
are ever in a state of suspense which makes all great conclusions impossible; the multiplicity of the facts of life, always present to consciousness,
restrains the free play of logic (except for that happy dreamer, the mathematician), and surrounds most of our pains and nearly all our pleasures
with infinite qualifications; we are tied down to a sober tameness. In our

dreams the fetters of civilisation are loosened, and we know the fearful joy of freedom.[21]

For the philosopher Henri Bergson (1859–1941), dreaming was, above all else, a form of disembodied thinking.

HENRI BERGSON, *DREAMS* (1914)

"You ask me what it is that I do when I dream? I will tell you what you do when you are awake. You take me, the me of dreams, me the totality of your past, and you force me, by making me smaller and smaller, to fit into the little circle that you trace around your present action. That is what it is to be awake. That is what it is to live the normal psychical life. It is to battle. It is to will. As for the dream, have you really any need that I should explain it? It is the state into which you naturally fall when you let yourself go, when you no longer have the power to concentrate yourself upon a single point, when you have ceased to will. What needs much more to be explained is the marvelous mechanism by which at any moment your will obtains instantly, and almost unconsciously, the concentration of all that you have within you upon one and the same point, the point that interests you. But to explain this is the task of normal psychology, of the psychology of waking, for willing and waking are one and the same thing."

This is what the dreaming ego would say. And it would tell us a great many other things still if we could let it talk freely. But let us sum up briefly the essential difference which separates a dream from the waking state. In the dream the same faculties are exercised as during waking, but they are in a state of tension in the one case, and of relaxation in the other. The dream consists of the entire mental life minus the tension, the effort and the bodily movement. We perceive still, we remember still, we reason still. All this can abound in the dream; for abundance, in the domain of the mind, does not mean effort. What requires an effort is the precision of adjustment. To connect the sound of a barking dog with the memory of a crowd that murmurs and shouts requires no effort. But in order that this sound should be perceived as the barking of a dog, a positive effort must be made. It is this force that the dreamer lacks. It is by that, and by that alone, that he is distinguished from the waking man.[22]

Through repeated self-suggestion, the British writer Mary Arnold-Forster (1861–1951) discovered, like others before her, that it was possible to "exercise a considerable degree of selection and control over our dreams." Through this acquired faculty, nightmares and dreams of distress could, she found, turn into cinematic entertainment.

MARY ARNOLD-FORSTER, *STUDIES IN DREAMS* (1921)

During the course of a long dream I had succeeded in tracing the existence of a complicated and dangerous plot against our country. The conspirators had turned upon me on discovering how much I knew. I was so closely followed, and my personal danger became so great, that the formula for breaking off a dream flashed into my mid and automatically gave me confidence; I remembered that I could make myself safe; but with the feeling of safety I also realised that if I were to wake my valuable knowledge of the dangerous conspiracy would be lost, for I realised that this was 'dream knowledge.' It was a dreadful dilemma—safety called me one way, but the conviction that my duty was to stay and frustrate the traitors was very strong. I feared that I should give way, and I knelt and prayed that I might have the courage not to seek safety by awakening, but to go on until I had done what was needed. I therefore did not wake; the dream continued. The arch-conspirator, a white-faced man in a bowler hat, had tracked me down to the building where I was concealed, and which by this time was surrounded; but all fear had departed, the comfortable feeling of great heroism, only fully enjoyed by those who feel themselves to be safe, was mine. It became a delightful dream of adventure, since the element of fear had gone.

This question of our power of control over dreams becomes a practical one, and one of serious importance, when we realise how closely it touches the health and happiness of our children; for the evil dreams that oppressed Charles Lamb's sensitive childhood are unhappily shared in more or less degree by many children, and are too often a cause of anguish to them. It would be a great gain if those who suffer thus could be helped to understand the nature of their troubles and to become to some extent masters of their dreams.[23]

But what about the world of the ordinary dreamer?

CALVIN HALL, "WHAT PEOPLE DREAM ABOUT" (1951)

What do people do in their dreams? We classified 2,668 actions in 1,000 dreams. By far the largest proportion (34 per cent) fall into the category of movement-walking, running, riding or some other gross change in bodily position. We found that, contrary to popular belief, falling or floating in dreams is not very common. After movement, the next most common activities were talking (11 per cent), sitting (7 per cent), watching (7 per cent), socializing (6 per cent), playing (5 per cent), manual work (4 per cent), thinking (4 per cent), striving (4 per cent), quarreling or fighting (3 per cent) and acquiring (3 per cent). From this it can be seen that passive or quiet activities occupy a large part of dreams, while manual activities are surprisingly infrequent. Such common waking occupations as typing, sewing, ironing and fixing things are not represented in these thousand dreams at all; cooking, cleaning house, making beds and washing dishes occur only once each. But strenuous recreational activities, such as swimming, diving, playing a game and dancing, are fairly frequent. In short, dreamers go places more than they do things; they play more than they work; their activities are more passive than active.[24]

Dreams go electric.

WILLIAM C. DEMENT, *SOME MUST WATCH WHILE SOME MUST SLEEP* (1974)

The discovery of the two kinds of sleep occurred almost accidentally at the University of Chicago. In 1952 Dr Kleitman became interested in the slow rolling of eye movements that accompany sleep onset and decided to look for these eye movements throughout the night to determine whether they were related to the depth or quality of sleep. Kleitman gave the assignment of watching eye movements to one of his graduate students in the department of physiology, Eugene Aserinsky. The young student soon noticed an entirely new kind of eye movement. At certain times during the night, the eyes began to dart about furiously beneath the closed lids. These unexpected episodes were startlingly different from the familiar slow, pendular movements that were the original object of the study.

Aserinsky was using the polygraph to monitor the subject, and the eye movements were actually discovered on the chart paper. It was not until we directly observed these movements in sleeping subjects that we could believe the spectacular inked out deviations. It would be different today to understand how skeptical we were. These eye movements, which had all the attributes of waking eye movements, had absolutely no business appearing in sleep. In those days, sleep was conceived of as a state of neural depression or inhibition—quiescence, rest. It was definitely not a condition in which the brain could be generating highly coordinated eye movements that were, in many instances, faster and sharper than the subject could execute while awake.

This was *the* breakthrough—the discovery that changed the course of sleep research from a relatively pedestrian inquiry into an intensely exciting endeavour pursued with great determination in laboratories and clinics all over the world. And there is nothing more exciting to a researcher than findings that are totally different from what he had expected.

Of course, the change in all our concepts of sleep didn't occur overnight. Having joined the research effort at this point as a sophomore medical student under Kleitman, I began to record the electroencephalograph and other physiological variables, along with eye movement activity. Since I didn't know what to expect, I kept my eyes glued to the moving chart paper all night long. After many nights, certain definite relationships were discernible in the enormous amounts of data. Rapid eye movements were always accompanied by very distinctive brain wave patterns, a change in breathing, and other striking departures from the normal, quiet sleep pattern. In addition, what we now call the basic ninety-minute sleep pattern began to emerge from the night-to-night variability.

As more and more physiological changes were discovered and described, we realized that sleep was *not* a quiet resting state that continued without variance as long as the subject was fortunate enough to remain asleep. No. For the first time we realized what has probably been true of man's sleep since he crawled out of the primordial slime. Man has *two* kinds of sleep. His nocturnal solitude contains two entirely *different* phenomena.[25]

Despite having little impact in understanding the psychology of dreaming, the new laboratory research continued to stimulate a wave of highly speculative theories on the nature and function of dreaming, including J. Allan Hobson's (1933–2021) Activation-Synthesis hypothesis, which began by attempting to understand the labile, scene-shifting nature of REM dreaming in terms of brain chemistry.

J. ALLAN HOBSON, *DREAM LIFE* (2011)

Whether a certain idea, emotion, fantasy, or dream scenario entered consciousness was, for us, not so much a question of freeing these psychic forms from repression as it was the result of a specific and selective sort of brain activation. Activation states differed most importantly according to the chemistry of the brain, not a shifting balance between such mental functions as superego, ego, and id. For . . . some other psychoanalytic critics of our work, the neurophysiological mechanisms we had uncovered only confirmed Freud's theory . . . the subcortical brain that was activated during REM was the id and the inactivated cortex was the superego. This Freudianization of physiology is understandable, but it does not deal with our main point. That point was our ability to explain the bizarreness of dreams without resorting to the Disguise-Censorship Model of psychoanalysis. For us, dreams were intrinsically bizarre, not bowdlerized translations of forbidden wishes.

As soon as we came up with the name Activation-Synthesis, McCarley and I knew we had formulated a model that was the antithesis of Freud's Disguise-Censorship Model. For Freud, there was activation too, but it was an unconscious wish that was activated because of a lapse in vigilance by the censor, an ego structure that normally protected consciousness from unacceptable id impulses. Enter disguise, the transformation of the latent dream content into the innocuous and seemingly jumbled manifest dream content. For us, the activation was physiological and selective: the limbic lobe and the posterolateral cortex were turned on, which led to automatic generation of emotions and associated imagery. The dream story was a synthesis of these disparate activations. That dream story was bizarre and confusing because the dorsolateral prefrontal cortex was deactivated and the entire forebrain was deprived of norepinephrine and serotonin, so that the forebrain was playing cards without a full deck. We later guessed that there must be some advantage to playing the game of life this way. But at the time we knew nothing of the importance of REM to temperature control, and we only dimly perceived the function of REM as a virtual reality generator for consciousness.[26]

But perhaps dreaming serves another purpose.

FRANCIS CRICK AND GRAEME MITCHISON, "FUNCTION OF DREAM SLEEP" (1983)

We propose here a new explanation for the function of REM sleep. The basis of our theory is the assumption that in viviparous mammals the cortical system (the cerebral cortex and some of its associated subcortical structures) can be regarded as a network of interconnected cells which can support a great variety of modes of mutual excitation. Such a system is likely to be subject to unwanted or 'parasitic' modes of behaviour, which arise as it is disturbed either by the growth of the brain or by the modifications produced by experience. We propose that such modes are detected and suppressed by a special mechanism which operates during REM sleep and has the character of an active process which is, loosely speaking, the opposite of learning. . . .

The mechanism we propose is based on the more or less random stimulation of the forebrain by the brain stem that will tend to excite the inappropriate modes of brain activity referred to earlier, and especially those which are too prone to be set off by random noise rather than by highly structured specific signals. We further postulate a reverse learning mechanism which will modify the cortex (for example, by altering the strengths of individual synapses) in such a way that this particular activity is less likely in the future. For example, if a synapse needs to be strengthened in order to remember something, then in reverse learning it would be weakened. Put more loosely, we suggest that in REM sleep we unlearn our unconscious dreams. "We dream in order to forget."[27]

The pioneer sleep scientist Michel Jouvet (1925–2017) proposed a parallel theory of dreaming as self-programming.

MICHEL JOUVET, *THE PARADOX OF SLEEP* (1999)

My thesis is that our dreams make each of us different because during them a repetitive program wipes out certain aspects of what we have learned, and may reinforce others if they are compatible with the "genetic program" of the dream . . . Iterative programming of the brain during paradoxical sleep would reinforce or obliterate the traces of epigenetic learning occurring during waking. Periodic dreaming would permit the repeated programming of unconscious reactions that are the basis of personality and individual differences in behavior in subjects exposed to the same environments. . . . As guardians and intermittent programmers of the hereditary part of our personality, it is possible that dreams also play a less conservational, predictive (Promethean) role. Indeed, thanks to the extraordinary diversity of possible connections in our brain as the basic circuits of our personality are programmed, an infinitely variable set of permutations could emerge—influenced by acquired experience—engendering the fantasies that make up dreams, or preparing new thought structures that will enable us to tackle new problems.[28]

David Foulkes (1935–) began his career as a sleep researcher at the University of Chicago, where his doctoral lab experiments established that William Dement and his colleagues had grossly underestimated the amount of mental activity that took place in non-REM Sleep. Two decades on, Foulkes went on to trace the development of dreaming in infants and children.

DAVID FOULKES, *CHILDREN'S DREAMING AND THE DEVELOPMENT OF CONSCIOUSNESS* (1999)

The popular belief is that young children and infants dream more or less the same way that we do—even if, necessarily, about different things.

Underlying this belief that dreaming is a given in human nature is the assumption that if one can see reality, one ought also be able to "see" imagined realities. I argue that dreaming is not like seeing, that its composition involves high-level thought processes rather than automatic perceptual ones. . . .

We now see that dreaming is not a "given" in human nature. Dreaming develops in predictable stages, and over a longer time span and far later in childhood than we may have imagined. The longitudinal study of children's dreaming identified the stages, and the cross-sectional study replicated the heart of this pattern: to begin with, static imaging; followed by kinematic imagery; and then by active participation in dream events by the self character, the last two typically occurring only sometime after age 7.

But these two studies have done something far more than merely chart the early development of human dreaming. They have furthered our understanding of dreaming itself. Specifically, they have shown us, in conjunction with the neuropsychological studies identifying dreaming with waking imaging, what sort of process dreaming is and what sort of process it isn't. In this regard, they provide the first really new insight into dreaming since the discovery in the 1950s of REM sleep, and the demonstration of its relation to particularly memorable dreaming.

We now see that dreaming is a very high-level cognitive process that has skill prerequisites lacking in infancy and which unfold slowly in the preschool and early primary-school years. We now know that dreaming is not an elementary and automatically engaged adjunct of waking perceptual processes, and that it cannot, therefore, be imagined to be present in every creature whose waking behavior is visually (or otherwise sensorially) guided. To dream, it isn't enough to be able to see. You have to be able to think in a certain way. Specifically, you have to be able, in your mind's eye, to simulate, at first momentarily and later in more extended episodes, a conscious reality that is not supported by current sense data and that you've never even experienced before.[29]

Might NREM sleep—most often approached as an impoverished sidekick to the REM dream, or as the basis for disorders of arousal—be more psychologically salient than the dream state itself?

DARIAN LEADER, *WHY CAN'T WE SLEEP?* (2019)

NREM almost always precedes REM in the sleep cycle, and its timing seems to be regulated by the amount of prior NREM. But should we then see REM as a kind of treatment or elaboration of what is being processed in NREM, an attempt to save NREM, or a temporary failure to sustain NREM? In other words, is REM a breakdown of some other state or process, a repair mechanism, or a logical development of this in its own right? . . .

Although NREM sleep is so often called 'quiet' here, it is possible that its deep, slow waves have been completely misunderstood. Several researchers have found 'GSR storms' during NREM, periods of sustained and intense arousal signalled by electrodermal activity. Likewise, the horrific night terrors of childhood and occasionally of adulthood emerge only from Stage 4 NREM, perhaps making the highly symbolised dream products associated with REM seem tame by comparison. The longer and deeper the slow-wave NREM sleep here, the more intense the night terror. This may also suggest that processes of symbolisation—that is, encryption and disguise—are denser in REM, and that hence, despite the rapid eye movements, breathing and cardiac accelerations, it is in fact the 'quieter' sleep.

The psychical activity of NREM sleep is much more difficult to remember than that of REM, which is one of the reasons why many researchers chose to believe that there was just less of it. If people who speak during NREM are woken up, they tend to have no memory of what they have said or even that they were speaking, yet if woken from REM sleep they may well remember dreams and fragments of speech. This may suggest, once again, that NREM sleep has more to hide, operating with intensities that, if rendered conscious, may be difficult to bear.[30]

2

CASTLES OF INDOLENCE: THE BORDERLANDS OF SLEEP

Not long after being elected president of the French Republic, in February 1920, Paul Deschanel found himself wandering at night along the railway line near the town of Montargis, seventy or so miles south of Paris. Not knowing where he was, the bloody-faced president, dressed in pajamas and socks, followed the tracks and was soon discovered by a railway worker, to whom he preceded to explain that his last memory was of boarding the Orient Express at the Gare de Lyon.

President Deschanel's much-lampooned exit from a moving train may well have been precipitated by a dose of chloral hydrate, a longstanding treatment for insomnia; or it might equally have been an episode of somnambulism, or confusional arousal, which a later generation of sleep scientists would classify as a non-rapid eye movement parasomnia. In either case, it confirmed that the borderlands between sleep and wakefulness remained ripe for clinical and experimental investigation, highlighting what little attention had been given to the drowsy and partial states of consciousness that lay at the flickering borders of sleep.

A century earlier, Henry Holland, physician extraordinary to Queen Victoria, had lamented that it was remarkable how little was known about sleep, despite "the perpetual experiment that life affords upon the subject."[1] Over the latter decades of the nineteenth century, a rising preoccupation with sleeplessness and cerebral fatigue brought new researchers into this virgin field, but for sleep to come within the purview of laboratory science required the development of instruments that permitted the physiology of the sleeping brain to be observed and recorded. And in 1878, the Turin-born physiologist Angelo Mosso did exactly this.

Using a specially devised "human circulation balance," Mosso was able to record the brain pulsations in a patient with a recent head wound, providing the first ever graphic illustration of brain activity during sleep. In the course of establishing that deep sleep was an active state, Mosso made a second important discovery: "At the slightest noise a wave of blood disturbed the surface of the brain. If the hospital clock struck the hour, or someone walked along the terrace, if I moved my chair, or wound up my watch, or if a patient coughed in the next room—everything, the slightest sound was accompanied by a marked alteration in the circulation of the brain, all immediately traced by the pen which the brain guided on the paper of my registering apparatus."[2] Even in deep sleep, the brain was at some level awake to external stimuli.

The discontinuities of sleep had, of course, long been attested outside the laboratory and clinic, and Mosso's experiment confirmed something that Aristotle, famed in his own lifetime as "the man who knew everything," had observed in his essays on sleep. There were, according to Aristotle, at least three ways in which consciousness might slip through the veil of sleep and create a state of half-somnolence. First, the sleeper could become aware of the fact that they were dreaming. Second, they might register external sights and sounds, such as the barking of dogs or crowing of cocks. Thirdly, and most dramatically, sleepers were known to "move in their sleep, and perform many waking acts" of which they retained no knowledge.[3]

Somnambulism would go on to become a widely debated legal and philosophical conundrum, the sleepwalker's "slumbery agitation" appearing to suggest the existence of a nocturnal self that could act independently of its daytime watchman. Most of the early encyclopedias included some mention of somnambulism—probably the most cited of all cases being that of a young ecclesiastic who fell into the habit of composing sermons and music while still asleep—but it is doubtful that all liminal antics purportedly undertaken by sleepwalkers arose within sleep. Almost any actions undertaken without apparent awareness, and without subsequent recall, were in this period described as instances of somnambulism, meaning that episodes of epilepsy, fugue, and hysterical automatism helped swell its conceptual ranks. Somnambulism could, moreover, be feigned by "sleeping preachers" whose impromptu sermons sometimes attracted the attention of pious followers.

A similar ambivalence surrounded daydreams and reveries, known principally as "abstraction," "fancy," and "wool-gathering" in the Anglophone world before the end of the seventeenth century. As Guy Claxton observes in *The Wayward Mind*, the medieval scholastics knew this drifting train of thought as "cogitation." Though regarded with suspicion by those who feared that its unbidden transports opened a doorway to the Devil, cogitation was often actively pursued by some monastics through the strategic use of half-sleep. "Thomas Aquinas, for instance, would have himself roused after a short sleep. And while still in that muzzy, in-between mode that modern psychologists called 'hypnagogia,' would lie prone on the ground to pray, and it would come to him what he was to write or dictate the following day."[4]

Over the coming centuries, the threshold between sleep and wakefulness continued to rouse fear and fascination. As the shadow of pathology slowly encroached upon the waking dreamer—framing their retreat into private fantasy as a symptom of monomania, or an analogue of insanity writ large—reverie's floating polyphony of thought became a full-blown literary motif, rebounding through the work of Montaigne, Rousseau, and De Quincey before surfacing in "psychological" novels which, in the words of Virginia Woolf, sought to capture "the flickerings of that innermost flame which flashes its messages through the brain."[5] And as novelists sought to capture this flame-flickering world, daydreaming began to be framed by educationalists, psychiatrists, and industrial psychologists as a form of mental absenteeism, a barrier to learning, mental health, and productivity.

In the meantime, sleep scientists were able to shine a light into the brain activity associated with daydreaming and other so-called parasomnias, demonstrating the variegated nature of sleep itself. Arguably the greatest breakthrough was down to the work of Alfred Lee Loomis, a Wall Street banker and science aficionado. An early adopter of the EEG machine at his private laboratory at Tuxedo Park, Loomis undertook a series of experiments in the 1930s that led to the identification of five states of sleep, each with its own signature EEG potentials.

Over the following decades, sleep researchers further explored the complex, cyclical nature of sleep using more sophisticated EEG technology to investigate its REM and non-REM stages. After Eugene Aserinsky and Nathaniel Kleitman's landmark research, four stages of quiescent non-REM sleep were found to account for around eighty percent of an average night's sleep, and it was within non-REM sleep that most disorders of arousal such as sleepwalking and confusional awakening appeared to occur. Yet even when supplemented by new brain imaging data, the sleep laboratory's neuroscientific paradigm still offered only a tiny glimpse of the biochemistry of sleep, and a still more glancing insight into the rich psychology of these borderland states of consciousness.

Thankfully, the borderlands of sleep were still patrolled by poets, novelists, psychologists, and psychoanalysts, all of whom were very much alive to the ways in which sleep and insomnia could, as Samuel Johnson put it, carry us into "a kind of twilight existence" somewhere "between dreaming and reasoning."[6] And while the tell-tale recordings of sleep spindles, k-complexes, and other micro-events promised to bring this hinterland into plain sight, there can be no denying that the microelectrode was still no match for a notebook or diary in probing the reveries, daydreams, half-awakenings, and other dimly lit vestibules through which we pass as we commute between sleep and wakefulness.

*

A century before the word "reverie" fell into popular usage, the French philosopher Michel de Montaigne (1533–1592) reflected on the fugitive thoughts that visited him, most often, in solitude.

MICHEL MONTAIGNE, "ON SOME VERSES OF VIRGIL" (1580)

But I am displeased with my mind for ordinarily producing its most profound and maddest fancies, and those I like the best, unexpectedly and when I am least looking for them; which suddenly vanish, having nothing to attach themselves to on the spot; on horseback, at table, in bed, but most only horseback, when my thoughts range most widely. In speech I am rather sensitively jealous of attention and silence if I am speaking in earnest; whoever interrupts me stops me. When I travel, the very necessity of the road cuts conversation short; besides I most often travel without company fit for these protracted discourses, whereby I get full leisure to commune with myself.

It turns out as with my dreams. While dreaming I recommend them to my memory (for I am apt to dream that I am dreaming); but the very next day I may well call to mind their coloring just as it was, whether gay, or sad, or strange, but as to what they were besides, the more I strain to find out, the more I plunge into oblivion. So of these chance thoughts that drop into my mind there remains in my memory only a vain notion, only as much as I need to make me rack my brains and fret in quest of them to no purpose.[7]

Early modern physicians and scholars often reported sleepwalkers who preached, composed sermons, played instruments, and fought imaginary assailants. Cervantes' Don Quixote, the greatest of all fictional daydreamers, headed the tilting charge of the sleepfighters.

MIGUEL DE CERVANTES, *DON QUIXOTE DE LA MANCHA* (1605)

There remained but little more of the novel to be read, when from the room, where Don Quixote lay, Sancho Panza came running out all in a fright, crying aloud: 'Run, sirs, quickly, and succour my master, who is over head and ears in the toughest and closest battle my eyes have ever beheld. As God shall save me, he has given the giant, that enemy of the princess Micomicona, such a stroke, that he has cut off his head close to his shoulders, as if it had been a turnip.' . . .

And, so saying, he went into the room, and the whole company after him; and they found Don Quixote in the strangest situation in the world. He was in his shirt, which was not quite long enough before to cover his thighs, and was six inches shorter behind: his legs were very long and lean, full of hair, and not over clean: he had on his head a little red cap, somewhat greasy, which belonged to the innkeeper. About his left arm he had twisted the bed-blanket (to which Sancho owed a grudge, and he very well knew why), and in his right hand he held his drawn sword, with which he was laying about him on all sides, and uttering words as if he had really been fighting with some giant: and the best of it was, his eyes were shut; for he was asleep, and dreaming that he was engaged in battle with the giant: for his imagination was so taken up with the adventure he had undertaken, that it made him dream he was already arrived at the kingdom of Micomicon, and already engaged in fight with his enemy; and, fancying he was cleaving the giant down, he had given the skins so many cuts, that the whole room was afloat with wine.[8]

The philosopher Ralph Cudworth (1617–1688) raised an all-too-common warning when he placed the waking dream under the specter of madness.

RALPH CUDWORTH, *A TREATISE CONCERNING ETERNAL AND IMMUTABLE MORALITY* (1731)

And the dreams that we have in our sleep are really the same kind of things with those imaginations that we have many times when we are awake, when the fancy, being not commanded or determined by the will, roves and wanders and runs at random, and spins out a long thread or concatenated series of imaginations or phantasms of corporeal things, quite different from those things which our outward senses at the same time take notice of. And some persons there are to whom these waking dreams are very ordinary and very familiar.

And there is little doubt to be made but if a man should suddenly fall asleep in the midst of one of these waking dreams when his fancy is roving and spinning out such a long series of imaginations, those very imaginations and phantasms would of course, *ipso facto*, become dreams and run on, and appear not as phantasms or imaginations only of things feigned or non-existent, but as perceptions of things really existent, that is as sensations. . . . And serious consideration hereof should make us very careful how we let the reins loose to that passive irrational party of our soul which knows no bounds or measures, least thereby we unawares precipitate and plunge ourselves headlong into the most sad and deplorable condition that is imaginable.[9]

The Enlightenment philosophe Jean-Jacques Rousseau (1712–1778) found immeasurable delight in the purdah of reverie.

JEAN-JACQUES ROUSSEAU, *REVERIES OF THE SOLITARY WALKER* (1782)

Even in our keenest pleasures there is scarcely a single moment of which the heart could truthfully say: 'Would that this moment could last for ever!' And how can we give the name of happiness to a fleeting state which leaves our hearts still empty and anxious, either regretting something that is past or desiring something that is yet to come? But if there is a state where the soul can find a resting-place secure enough to establish itself and concentrate its entire being there, with no need to remember the past or reach into the future, where time is nothing to it, where the present runs on indefinitely but this duration goes unnoticed, with no sign of the passing of time, and no other feeling of deprivation or enjoyment, pleasure or pain, desire or fear than the simple feeling of existence, a feeling that fills our soul entirely, as long as this state lasts, we can call ourselves happy, not with a poor, incomplete and relative happiness such as we find in the pleasures of life, but with a sufficient, complete and perfect happiness which leaves no emptiness to be filled in the soul. Such is the state which I often experienced on the Island of Saint-Pierre in my solitary reveries, whether I lay in a boat and drifted where the water carried me, or sat by the shores of the stormy lake, or elsewhere, on the banks of a lovely river or a stream murmuring over the stones. What is the source of our happiness in such a state? Nothing external to us, nothing apart from ourselves and our own existence; as long as this state lasts we are self-sufficient like God.[10]

"Natural" and "artificial" somnambulism were subjects of great interest to the Romantics, but the essayist William Hazlitt (1778–1830) had a more intimate perspective on this "state of half-perception" than most of his contemporaries.

WILLIAM HAZLITT, "ON DREAMS" (1826)

I myself am (or used some time ago to be) a sleep-walker; and know how the thing is. In this sort of disturbed, unsound sleep, the eyes are not closed, and are attracted by the light. I used to get up and go towards the

window, and make violent efforts to throw it open. The air in some measure revived me, or I might have tried to fling myself out. I saw objects
indistinctly, the houses, for instance, facing me on the opposite side of
the street; but still it was some time before I could recognise them or
recollect where I was: that is, I was still asleep, and the dimness of my
senses (as far as it prevailed) was occasioned by the greater numbness of
my memory. This phenomenon is not astonishing, unless we choose in
all such cases to put the cart before the horse. For in fact, it is the mind
that sleeps, and the senses (so to speak) only follow the example. The
mind dozes, and the eyelids close in consequence: we do not go to sleep
because we shut our eyes. I can, however, speak to the fact of the eyes
being open when their sense is shut; or rather, when we are unable to
draw just inferences from it. It is generally in the night-time, indeed, or
in a strange place, that the circumstance happens; but as soon as the light
dawns on the recollection, the obscurity and perplexity of the senses clear
up. The external impression is made before, much in the same manner as
it is after we are awake; but it does not lead to the usual train of associations connected with that impression; e.g., the name of the street or town
where we are, who lives at the opposite house, how we came to sleep in
the room where we are, &c.; all which are ideas belonging to our waking
experience, and are at this time cut off or greatly disturbed by sleep. It is
just the same as when persons recover from a swoon, and fix their eyes
unconsciously on those about them, for a considerable time before they
recollect where they are. . . . The stupor is general: the faculty of thought
itself is impaired; and whatever ideas we have, instead of being confined
to any particular faculty or the impressions of any one sense, and invigo-
rated thereby, float at random from object to object, from one class of
impressions to another, without coherence or control. The conscious
or connecting link between our ideas, which forms them into separate
groups or compares different parts and views of a subject together, seems
to be that which is principally wanting in sleep; so that any idea that
presents itself in this anarchy of the mind is lord of the ascendant for
the moment, and is driven out by the next straggling notion that comes
across it. The bundles of thought are, as it were, untied, loosened from a
common centre, and drift along the stream of fancy as it happens. Hence
the confusion (not the concentration of the faculties) that continually

takes place in this state of half-perception. The mind takes in but one thing at a time, but one part of a subject, and therefore cannot correct its sudden and heterogeneous transitions from one momentary impression to another by a larger grasp of understanding.[11]

How much of our lives do we pass in a "Brown Study" of mental indolence, asks the radical philosopher William Godwin (1756–1836).

WILLIAM GODWIN, *THOUGHTS ON MAN* (1831)

But the lives of all men, the wise, and those whom by way of contrast we are accustomed to call the dull, are divided between animation and comparative vacancy; and many a man, who by the bursts of his genius has astonished the world, and commanded the veneration of successive ages, has spent a period of time equal to that occupied by a walk from Temple-Bar to Hyde-Park-Corner, in a state of mind as idle, and as little affording materials for recollection, as the dullest man that ever breathed the vital air.

The two states of man which are here attempted to be distinguished, are, first, that in which reason is said to fill her throne, in which will prevails, and directs the powers of mind or of bodily action in one channel or another; and, secondly, that in which these faculties, tired of for ever exercising their prerogatives, or, being awakened as it were from sleep, and having not yet assumed them, abandon the helm, even as a mariner might be supposed to do, in a wide sea, and in a time when no disaster could be apprehended, and leave the vessel of the mind to drift, exactly as chance might direct.

To describe this last state of mind I know not a better term that can be chosen, than that of reverie. It is of the nature of what I have seen denominated BROWN STUDY a species of dozing and drowsiness, in which all men spend a portion of the waking part of every day of their lives. Every man must be conscious of passing minutes, perhaps hours of the day, particularly when engaged in exercise in the open air, in this species of neutrality and eviration. It is often not unpleasant at the time, and leaves no sinking of the spirits behind. It is probably of a salutary nature, and may be among the means, in a certain degree beneficial like sleep, by which the machine is restored, and the man comes forth from its discipline reinvigorated, and afresh capable of his active duties.

This condition of our nature has considerably less vitality in it, than we experience in a complete and perfect dream. In dreaming we are often conscious of lively impressions, of a busy scene, and of objects and feelings succeeding each other with rapidity. We sometimes imagine ourselves earnestly speaking: and the topics we treat, and the words we employ, are supplied to us with extraordinary fluency. But the sort of vacancy and inoccupation of which I here treat, has a greater resemblance to the state of mind, without distinct and clearly unfolded ideas, which we experience before we sink into sleep. The mind is in reality in a condition, more properly accessible to feeling and capable of thought, than actually in the exercise of either the one or the other. We are conscious of existence and of little more. We move our legs, and continue in a peripatetic state; for the man who has gone out of his house with a purpose to walk, exercises the power of volition when he sets out, but proceeds in his motion by a semi-voluntary act, by a sort of vis inertiae, which will not cease to operate without an express reason for doing so, and advances a thousand steps without distinctly willing any but the first. When it is necessary to turn to the right or the left, or to choose between any two directions on which he is called upon to decide, his mind is so far brought into action as the case may expressly require, and no further.

I have here instanced in the case of the peripatetic: but of how many classes and occupations of human life may not the same thing be affirmed? It happens to the equestrian, as well as to him that walks on foot. It occurs to him who cultivates the fruits of the earth, and to him who is occupied in any of the thousand manufactures which are the result of human ingenuity. It happens to the soldier in his march, and to the mariner on board his vessel. It attends the individuals of the female sex through all their diversified modes of industry, the laundress, the housemaid, the sempstress, the netter of purses, the knotter of fringe, and the worker in tambour, tapestry and embroidery. In all, the limbs or the fingers are employed mechanically; the attention of the mind is only required at intervals; and the thoughts remain for the most part in a state of non-excitation and repose.

It is a curious question, but extremely difficult of solution, what portion of the day of every human creature must necessarily be spent in this sort of intellectual indolence. In the lower classes of society its empire is certainly very great; its influence is extensive over a large portion of the opulent and luxurious; it is least among those who are intrusted in the more serious affairs of mankind, and among the literary and the learned, those who waste their lives, and consume the midnight-oil, in the search after knowledge.[12]

The alienist Jacque-Joseph Moreau (1804–1884), the first physician to systematically study the psychological effects of cannabis, found that hashish intoxication produced a special kind of waking dream—"a composite state of madness and reason" that could be used to better understand the vagaries of insanity.

JACQUES-JOSEPH MOREAU, *HASHISH AND MENTAL ILLNESS* (1845)

The action of hashish weakens the will—the mental power that rules ideas and associates and connects them together. Memory and imagination become dominant; present things become foreign to us, and we are concerned entirely with things of the past and the future. . . . There occurs an uninterrupted succession of true and false ideas, of dreams and realities, which constitutes a composite state of madness and reason and makes a person seem mad and rational at the same time.

As the disorder of the faculties increases, as the storm which perturbs the faculties becomes more violent, consciousness is carried away by the whirlwind and is a toy of one's dreams. *Lucid* moments are increasingly brief. Mental activity seems to withdraw and to restrict itself entirely to the brain. We abandon ourselves to our inner feelings; our eyes and our ears do not cease to function, but they admit only those impressions supplied by memory or imagination. In other words, *we fall asleep while dreaming.*

But then, as if consciousness could never be completely extinguished, here is what happens—I will be better understood if I recall a fact well known to those who dream often without ceasing to sleep—we are sometimes aware that we are dreaming. Better than that, when the dream pleases us, we are afraid to awaken. We force ourselves to prolong the dream, and when we sense that it is going to end, we say to ourselves: 'Why is all that only a dream?' This is exactly the same state that is experienced by a person under the influence of hashish in its most potent form.[13]

While Freud famously observed that "Hysterics suffer mainly from reminiscences," his contemporary Pierre Janet (1859–1947) found that their "ceaseless reveries" led to the tyranny of "fixed ideas."

PIERRE JANET, *THE MENTAL STATE OF HYSTERICALS* (1892)

But there is a manifestation of intellectual automatism practically more important than all the others, namely, the tendency to ceaseless reverie. Hystericals are not content to dream constantly at night; they dream all day long. Whether they walk, or work, or sew, their minds are never wholly occupied with what they are doing. They carry on in their heads an interminable story which unrolls before them or is inwardly conceived. You think Justine is listening to you while you talk to her? No, for all at once you hear her murmur low: "No, sir, you are a savage." When you shake her, she will excuse herself and tell you that she was thinking of a horrible policeman who was going to carry off a little dog to the pound. Bertha makes us constantly repeat to her what we have said, and adds: "It is not my fault; I was no longer listening; I fancied I was disinterring my mother and hovered with her in the clouds." Her eyes seem to read a book, but her face tells another story: "When I shall die, I shall have a small white bouquet on my small tomb, . . . flowers speak, they are very gentle per-sons; I shall talk to them, for they are not bad. . . ." And here we can see her tears flow over a book which is very entertaining and gives not the least occasion for her grief. It is certain that it is not easy to understand what such an one reads when such stories are told.

These reveries sometimes have no development; they are variable, incoherent images, which pass before us like the colours of a kaleidoscope, though they have often a certain vague unity about them. It is always the same monotonous story which the patient resumes at the point where she has been interrupted, or unceasingly begins over again. Be that story cheerful or painful, it does not matter; it becomes pleasant to these weary minds, be-cause it is an easy reverie: "How unhappy I am! . . . The idea is frightful, but it lulls me, and I am ready to defend it against any one who should want to take it from me; let me think of my little tomb; it gives me so much pleasure. . . ." When reveries get to be systematised in this way, they become more dangerous and are soon transformed into fixed ideas.[14]

The pioneer sleep researcher Marie de Manaciene (1841–1903) undertook the first laboratory studies of a peculiar state of "half-awakening" now termed "confusional arousal."

MARIE DE MANACIENE, *SLEEP: ITS PHYSIOLOGY, PATHOLOGY, HYGIENE, AND PSYCHOLOGY* (1897)

There are many persons who present this state when they are awakened in the middle of a profound sleep, but only for a short time. When such subjects are thus waked up they do not at once regain their usual consciousness; they appear to be confused and not know where they really are and what is wanted of them; but this state lasts only a few moments. Experience soon showed me that one cannot use cases of half-awakening for experiments of any kind, save perhaps for suggestions of fear, when they present a duration of less than fifteen seconds. And even the suggestions of fear are very far from being always successful in those cases, because persons in whom the state of half-awaking is very short-lived are not easily brought under suggestions even of emotional kind, so that when you shout at them, 'We are burning,' or 'House on fire,' instead of getting frightened at once in a reflex manner, they wake up instantly and completely, and with the air of a conscious observer ask, 'Where is the fire?' and 'What is burning?' Even in cases when one is successful enough to suggest the emotion of fear to those persons with a short-lived half-awakening, it lasts too short a time. This made me decide to count as cases of half-awakening only those in which the state has a duration of more than fifteen seconds. My observations showed me that this state of half-awakening can last from fifteen seconds to six minutes; in some cases, the half-awakening last even more than six minutes, but cases of this kind are to be considered wholly abnormal, and they must inspire us with anxious doubts as to the mental health of persons presenting with unusually long periods of half-awakening.[15]

In one of the earliest questionnaire studies of daydreaming, Clark University psychologist Theodate L. Smith (1860–1914) asked more than a thousand school children to write about their day dreams.

THEODATE L. SMITH, "THE PSYCHOLOGY OF DAY DREAMS" (1904)

For girls from eight to ten, the fairy tale form of day dream predominates over all others. It appears to be a mental device for compassing all desires, and actual experiences and possibilities are often mingled indiscriminately with the wildest impossibilities. Nearly all dream of being rich and having every desire gratified and the dream of being a princess and living in a palace "with a piano in every room" and having unlimited silk dresses and jewels may be mixed with the wish "to have enough good food every day." The deus ex machina in these dreams is most frequently a fairy godmother, though wishing caps, a magic lamp or ring also figure. With boys of this age, the fairy story dream is less common and the form differs from that among girls. An interesting example of this occurred in a grade where the children were evidently all familiar with the story of Aladdin's lamp and the magic carpet. Nearly all of these had day dreams of flying or being transported through the air. Nearly all the girls had preserved the original forms of the stories with slight alterations, but the boys dreamed of all sorts of wonderful flying machines, sometimes mentioning the rate per hour, of trips in a balloon or by means of mechanical wings, of which they were in some cases the inventors. The desire for riches, while quite as wide-spread among boys as among girls, seems to demand a more logical explanation of its origin than that furnished by a fairy godmother. Dreams of finding money in amounts varying from fifty cents to five million dollars occur or the dream may be projected into the future and acquiring a fortune by possible or impossible means may be imagined, but however improbable the dream there is usually an attempt at logical consistency. . . .

From the age of twelve the influence of books upon the content of the day dream becomes increasingly important. With the less imaginative the dream may be merely a reproduction, with slight alterations, of some book recently read, but in other cases the book simply furnishes the raw material out of which the fabric of the dream is woven. Girls put themselves in the place of their favorite heroines and adapt the material of romance, poetry or travels to their own uses. Their ideals of life are affected by what they read. Some of these dreams of the future are visions of beautiful and useful womanhood, but the trail of the Elsie books, with their morbid religiosity, and the influence of the Duchess and Rhoda Broughton is evident with unfortunate frequency. Boys dream of fighting Indians, having hair breadth adventures on land and sea, being cowboys, pirates, brigands or national heroes as the case may be. Detective stories seem to acquire a peculiar charm at about the age of fourteen.[16]

The Viennese psychoanalyst Herbert Silberer (1882–1923), best known for his theories of symbol formation, found that his daydreams carried him into a kind of rudimentary "picture-thinking."

HERBERT SILBERER, *REPORT ON A METHOD OF ELICITING AND OBSERVING CERTAIN SYMBOLIC-HALLUCINATION PHENOMENA* (1909)

One afternoon (after lunch) I was lying on my couch. Though extremely sleepy, I forced myself to think through a problem of philosophy, which was to compare the views of Kant and Schopenhauer concerning time. In my drowsiness I was unable to sustain their ideas side by side. After several unsuccessful attempts, I once more fixed Kant's argument in my mind as firmly as I could and turned my attention to Schopenhauer's. But when I tried to reach back to Kant, his argument was gone again, and beyond recovery. The futile effort to find the Kant record which was somehow misplaced in my mind suddenly represented itself to me as I was lying there with my eyes closed, as in a dream as a perceptual symbol: I am asking a morose secretary for some information; he is leaning over his desk and disregards me entirely; he straightens up for a moment to give me an unfriendly and rejecting look.

The vividness of the unexpected phenomenon surprised, indeed almost frightened me. I was impressed by the appropriateness of this unconsciously selected symbol. I sensed what might be the conditions for the occurrence of such phenomena, and decided to be on the lookout for them and even to attempt to elicit them. In the beginning I hoped that this would yield a key to "natural symbolism." From its relation to art-symbolism I expected and still expect the clarification of many psychological, characterological, and aesthetic issues. . . .

What had happened? In my drowsiness my abstract ideas were, without any conscious interference, replaced by a perceptual picture by a symbol. My abstract chain of thoughts was hampered; I was too tired to go on thinking in that form; the perceptual picture emerged as an "easier" form of thought. It afforded an appreciable relief, comparable to the one experienced when sitting down after a strenuous walk, It appears to follow as a corollary that such "picture-thinking" requires less effort than the usual kind. The tired consciousness, not having at its disposal the energy necessary for normal thinking, switches to an easier form of functioning.[17]

"Surrealism is based on the belief in the superior reality of certain forms of association, heretofore neglected, in the omnipotence of the dream, and in the disinterested play of thought," wrote André Breton, co-founder and leading theorist of the Surrealist movement. The origins of this trickster-ish credo can be traced in some degree back to Breton's youthful experiences of hypnagogia.

ANDRÉ BRETON, *WHAT IS SURREALISM?* (1934)

It was in 1919, in complete solitude and at the approach of sleep, that my attention was arrested by sentences, more or less complete, which became perceptible to my mind without my being able to discover (even by meticulous analysis) any possible previous volitional effort. One evening in particular, as I was about to fall asleep, I became aware of a sentence articulated clearly to a point excluding all possibility of alteration and stripped of all quality of vocal sound; a curious sort of sentence which came to me bearing—in sober truth—not a trace of any relation whatever to any incidents I may at that time have been involved in; an insistent sentence, it seemed to me; a sentence, I might say, that *knocked at the window.*

I was prepared to divert my attention from it when the organic character of the sentence detained me. I was really bewildered. Unfortunately, I am unable at this distance to remember the exact sentence, but it ran something like this: "A man is cut in half by the window." What made it clearer was the fact that it was accompanied by a feeble visual representation of a man in the process of walking, but cloven, at half his height, by a window perpendicular to the axis of his body. Definitely, there was the form, re-erected against space, of a man leaning out of a window. But, with the window following the man's locomotion, I understood that I was dealing with an image of great rarity. Instantly the idea came to me to use it as material for poetic construction. I no sooner had invested it with that quality than it had given place to a succession of all but intermittent sentences which left me no less astonished, but in a state, I would say, of extreme detachment.[18]

Romanian-born French philosopher Emil Cioran (1911–1995) spent much of his life estranged from sleep. Insomnia was, he contended, "so full and so vacant that it suggests itself as a rival of time."

E. M. CIORAN, *ON THE HEIGHTS OF DESPAIR* (1934)

Whoever said that sleep is the equivalent of hope had a penetrating intuition of the frightening importance not only of sleep but also of insomnia! The importance of insomnia is so colossal that I am tempted to define man as the animal who cannot sleep. Why call him a rational animal when other animals are equally reasonable? But there is not another animal in the entire creation that wants to sleep yet cannot. Sleep is forgetfulness: life's drama, its complications and obsessions vanish completely, and every awakening is a new beginning, a new hope. Life thus maintains a pleasant discontinuity, the illusion of permanent regeneration. Insomnia, on the other hand, gives birth to a feeling of irrevocable sadness, despair, and agony. The healthy man—the animal—only dabbles in insomnia: he knows nothing of those who would give a kingdom for an hour of unconscious sleep, those as terrified by the sight of a bed as they would be of a torture rack. There is a close link between insomnia and despair. The loss of hope comes with the loss of sleep. The difference between paradise and hell: you can always sleep in paradise, never in hell. God punished man by taking away sleep and giving him knowledge. Isn't deprivation of sleep one of the most cruel tortures practiced in prisons? Madmen suffer a lot from insomnia; hence their depressions, their disgust with life, and their suicidal impulses. Isn't the sensation, typical of wakeful hallucinations, of diving into an abyss, a form of madness? Those who commit suicide by throwing themselves from bridges into rivers or from high rooftops onto pavements must be motivated by a blind desire to fall and the dizzying attraction of abysmal depths.[19]

As a novelist and critic, Vladimir Nabokov (1899–1977) prized stylistic control and exacting prose, which may explain why he was far from impressed by the "trivial" and "grotesque" theatre he often entered on the edge of sleep.

VLADIMIR NABOKOV, *SPEAK, MEMORY* (1947)

Just before falling asleep, I often become aware of a kind of one-sided conversation going on in an adjacent section of my mind, quite independently from the actual trend of my thoughts. It is a neutral, detached, anonymous voice, which I catch saying words of no importance to me whatever—an English or a Russian sentence, not even addressed to me, and so trivial that I hardly dare give samples, lest the flatness I wish to convey be marred by a molehill of sense. This silly phenomenon seems to be the auditory counterpart of certain praedormitary visions, which I also know well. What I mean is not the bright mental image (as, for instance, the face of a beloved parent long dead) conjured up by a wing-stroke of the will; that is one of the bravest movements a human spirit can make. Nor am I alluding to the so-called muscae volitantes—shadows cast upon the retinal rods by motes in the vitreous humor, which are seen as transparent threads drifting across the visual field. Perhaps nearer to the hypnagogic mirages I am thinking of is the colored spot, the stab of an afterimage, with which the lamp one had just turned off wounds the palpebral night. However, a shock of this sort is not really a necessary starting point for the slow, steady development of the visions that pass before my closed eyes. They come and go, without the drowsy observer's participation, but are essentially different from dream pictures for he is still master of his senses. They are often grotesque. I am pestered by roguish profiles, by some coarse-featured and florid dwarf with a swelling nostril or ear. At times, however, my photisms take on a rather soothing flou quality, and then I see—projected, as it were, upon the inside of the eyelid—gray figures walking between beehives, or small black parrots gradually vanishing among mountain snows, or a mauve remoteness melting beyond moving masts.[20]

First appearing in the pages of the New Yorker, Walter Mitty became an eponym for daydreamers and fantasists who took refuge in imaginary heroics.

JAMES THURBER, *THE SECRET LIFE OF WALTER MITTY* (1939)

"We're going through!" The Commander's voice was like thin ice breaking. He wore his full-dress uniform, with the heavily braided white cap pulled down rakishly over one cold gray eye. "We can't make it, sir. It's spoiling for a hurricane, if you ask me." "I'm not asking you, Lieutenant Berg," said the Commander. "Throw on the power lights! Rev her up to 8,500! We're going through!" The pounding of the cylinders increased: ta-pocketa-pocketa-pocketa *pocketa-pocketa*. The Commander stared at the ice forming on the pilot window. He walked over and twisted a row of complicated dials. "Switch on No. 8 auxiliary!" he shouted. "Switch on No. 8 auxiliary!" repeated Lieutenant Berg. "Full strength in No. 3 turret!" shouted the Commander. "Full strength in No. 3 turret!" The crew, bending to their various tasks in the huge, hurtling eight-engine Navy hydroplane, looked at each other and grinned. "The Old Man'll get us through," they said to one another. "The Old Man ain't afraid of Hell!" . . .

"Not so fast! You're driving too fast!" said Mrs. Mitty. "What are you driving so fast for?"

"Hmm?" said Walter Mitty. He looked at his wife, in the seat beside him, with shocked astonishment. She seemed grossly unfamiliar, like a strange woman who had yelled at him in a crowd. "You were up to fifty-five," she said. "You know I don't like to go more than forty. You were up to fifty-five." Walter Mitty drove on toward Waterbury in silence, the roaring of the SN202 through the worst storm in twenty years of Navy flying fading in the remote, intimate airways of his mind. "You're tensed up again," said Mrs. Mitty. "It's one of your days. I wish you'd let Dr. Renshaw look you over."[21]

The American-born neurophysiologist and cybernetician, William Grey Walter (1910–1977), creator of the first generation of autonomous robots, was at the vanguard of laboratory research on the electrical signature of sleep.

WILLIAM GREY WALTER, *THE LIVING BRAIN* (1953)

When a subject begins to feel drowsy, the first sign is usually a *reduction* in alpha rhythms and their gradual replacement by something more like the theta activity, mainly at the back and sides of the brain but spreading occasionally to all regions. At this stage the subject is easily roused and not yet really dozing; it is the beginning of that delicious state when consciousness is consciously waning. If the subject is trying *not* to shut the eyes, he will find himself seeing double; if reading, he will go over a paragraph time and again without getting its meaning. Some have called this state 'floating', a metaphor that well describes the freedom from bodily care which it promotes. This is the clue to understanding the nature of this stage of torpor; it is as though the whole body were being shut out from mental view, as the outside world is shut out by closing the eyes. Here of course there is no mechanical or anatomical shutter for the muscles and joints that can deceive the brain into neglect of its supervisory commission; there is, however, an elaborate mechanism in the base and stem of the brain that, when fatigue or custom dictates, inexorably weakens the significance of the incoming flood of sense data. This is the moment when the exhausted driver begins to persuade himself that the road is so straight that he can safely drive a while with his eyes shut—a comforting thought though it be his last.

At this stage the brain can still respond electrically and functionally to incoming signals, but the electrical response begins to show a prominent slow component, and the spread of the faster excitatory effects is more limited and more transient. This is the objective sign of what we called above a *weakening of the significance* of the signals. It is not the direct transmission of the nerve impulses to the projection areas that is inhibited, but rather their dissemination—the first two operations of learning—that is attenuated. The reader's eyes can follow the lines of print but the meaning of the words escapes him and, merging into private fancy, his bedtime story becomes a dream.[22]

Drawing on laboratory studies of sleep deprivation, American clinicians suggested that psychotic symptoms emerged at around a hundred hours of wakefulness.

LOUIS J. WEST ET AL., "THE PSYCHOSIS OF SLEEP DEPRIVATION" (1962)

By day, attention span shortens, intrusive thoughts become more and more prominent, and fleeting hallucinations of two types begin to occur. The first type, perhaps entoptic in origin, is of high-frequency rhythmical movements of actually stationary objects and, then, of patterned forms: grillwork, networks, reticulations, filigrees, laceworks, cobwebs, rippling water, and geometric designs. The second type resembles hypnogogic phenomena, and consists of brief dreamlike experiences in which total situations are perceived and reacted to.

With all of these phenomena, there is a growing sense of fatigue, weariness, drowsiness, disinterest in the outside world, and a tendency to withdraw. . . . As the period of sleep deprivation goes on, periods of overt confusion and clouding of consciousness are seen. Disorientation becomes more frequent and prolonged; first for time, then for place, then for person and, finally, for self. Gross delusional thinking, usually paranoid, becomes increasingly prominent. By night the subject gives a picture resembling a case of toxic delirium, with lucid intervals growing fewer and shorter. By day the picture is more like a case of paranoid schizophrenia, perhaps of the reactive type, with a progressively increasing intensity of psychopathology including suspiciousness, emotional lability, and delusions of reference, grandeur, and persecution.

The total impression is of a progressive disorganization of ego structure, modified by the influence of a 24-hour cycle and, also, by other, briefer periodic influences, perhaps in the range of 90 to 120 min. This is particularly notice-able at night, when a series of gross psychotic episodes may be seen to occur, for all the world like dream episodes during sleep, and very likely reflecting the same basic neurobiological periodicity or rhythm that underlies dreaming.[23]

In his three-volume magus opus The Principle of Hope, Frankfurt School philosopher Ernst Bloch (1885–1977) praised the utopian ferment of daydreams.

ERNST BLOCH, *THE PRINCIPLE OF HOPE* (1959)

In contrast to the nocturnal dream, that of the daytime sketches freely chosen and repeatable figures in the air, it can rant and rave, but also brood and plan. It gives free play to its thoughts in an indolent fashion (which can, however, be closely related to the Muse and to Minerva), political, artistic, scientific thoughts. The daydream can furnish inspirations which do not require interpreting, but working out, it builds castles in the air as blueprints too, and not always just fictitious ones. Even in caricature, the daydreamer is presented in a different light than the dreamer: he is then Johnnie Head-in-the-air, and thus by no means the sleeper at night with his eyes closed. . . . Psychoanalysis of course, which judges all dreams only as roads to what has been repressed, and only knows reality as that of bourgeois society and its existing world, consistently prefers to label daydreams as a mere stepping-stone to nocturnal ones. In any case, the poet equipped with daydreams is for the bourgeois only the hare who sleeps with his eyes open, and this in bourgeois everyday life which sees and employs itself as the touchstone of all reality. But if this touchstone is challenged even for the world of consciousness, if even the nocturnal wishful dream is only seen as a dislocated and not entirely homogeneous component in the vast field of a still open world and its consciousness, then the daydream is not a stepping-stone to the night-dream and is not disposed of by the latter. Not even with respect to its clinical content, let alone its artistic, pre-appearing, frontlike anticipatory content. For night-dreams mostly cannibalize the former life of the drives, they feed on past if not archaic image-material, and nothing new happens under their bare moon. So it would be absurd to take daydreams: as those presentiments of the imagination which from time immemorial have of course been called dreams but also forerunners and anticipations, and to subsume them under or even subordinate them to the night-dreams. The castle in the air is not a stepping-stone to the nocturnal labyrinth, if anything, the nocturnal labyrinths lie like cellars beneath the daytime castle in the air. And what of the equality of imaginary happiness which both are said to share, as a 'restoration of the independence of pleasure-gaining from the consent of reality'? More than one daydream before now has, with sufficient vigour and experience, remodelled reality to make it give this consent; whereas Morpheus only has the arms in which we rest. Thus the daydream requires specific evaluation of its own, since it enters and unlocks a very different region altogether. It ranges from the waking dream of a comfortable, silly, crude, escapist, devious and paralysing kind, to the responsible kind, the kind actively and acutely deployed in the matter-in-hand, and the shaped kind in art. Above all, it is clear that 'reverie', unlike the usual nocturnal 'dream', can possibly contain marrow and, instead of the idleness or even the self-enervation which certainly are to be found here, a tireless incentive towards the actual attainment of what it visualizes.[24]

Gaston Bachelard (1884–1962), France's foremost philosopher of the imagination, believed that the purest recollection of childhood was to be found in the "dreamed memories" of reverie.

GASTON BACHELARD, *THE POETICS OF REVERIE* (1960)

The being of reverie crosses all the ages of man from childhood to old age without growing old. And that is why one feels a sort of redoubling of reverie late in life when he tries to bring the reveries of childhood back to life.

This reinforcement of reverie, this deepening of reverie which we feel when we dream of our childhood explains that, in all reverie, even that which takes us into the contemplation of a great beauty of the world, we soon find ourselves on the slope of memories; imperceptibly, we are being led back to old reveries, suddenly so old that we no longer think of dating them. A glimmer of eternity descends upon the beauty of the world. We are standing before a great lake whose name is familiar to geographers, high in the mountains, and suddenly we are returning to a distant past. We dream while remembering. We remember while dreaming. Our memories bring us back to a simple river which reflects a sky leaning upon hills. But the hill gets bigger and the loop of the river broadens. The little becomes big. The world of childhood reverie is as big, bigger than the world offered to today's reverie. From poetic reverie, inspired by some great spectacle of the world to childhood reverie, there is a commerce of grandeur. And that is why childhood is at the origin of the greatest landscapes. Our childhood solitudes have given us the primitive immensities.

By dreaming on childhood, we return to the lair of reveries, to the reveries which have opened up the world to us. It is reverie which makes us the first inhabitant of the world of solitude.[25]

3

OPERATION TRANCE: FROM MESMERISM TO MIND CONTROL

From vital spark to animal heat, phlogiston to subtle spirit, eighteenth-century science was no stranger to factitious elements and forces that were believed to sustain the natural world. Franz Anton Mesmer's speculations on animal magnetism, first outlined in his 1779 *Mémoire sur la découverte du magnétisme animal*, were at one with the era's flurry of electrical and chemical speculations, but the legacy of the cures that the Viennese physician-cum-showman effected through ethereal currents proved altogether more far-reaching. Finding a place in medicine, philosophy and Spiritualism, animal magnetism, also known as mesmerism, gained traction as both a socio-religious credo and popular entertainment before providing the impetus for clinical and experimental studies of hypnosis. Without Mesmer's questionable panacea, many of the central tenets of scientific and lay psychology, psychiatry, and psychoanalysis would be missing their founding legends.

Mesmer's bid for scientific glory began to unravel in 1785, when two committees were appointed to investigate this most alternative of medicines. Working in conjunction with the University of Paris Faculty of Medicine, the committee of the Royal Academy of Sciences reported that magnetism without the willing imagination was redundant. The screams, convulsions, hysterical laughter—all the effects that worked their way into the healing "crisis" that Mesmer and his followers orchestrated— were ascribed to the workings of excitement, imagination, and imitation.

As Mesmer's medical heterodoxy travelled through Europe and America, the trance state became a psycho-physical Rosetta stone: a tableau whose amorphous constellation of symptoms were variously mapped, ordered, and interpreted. One early theorist, Carl Kluge, identified six stages of magnetic trance, moving from physical flushes through to clairvoyance. Half a century later, the pioneering neurologist Jean-Marie Charcot proposed that there were only three stages of entrancement—the lethargic, cataleptic, and somnambulistic—all of which flourished on the pathological soil of hysteria.

By the 1890s, the suggestion theory of hypnosis prevailed, finding that the trance state had, as William James observed in the *Principles of Psychology*, "no particular outward symptoms of its own."[1] Amnesia, paralysis, analgesia, compliance, and hyperaesthesia—every effect that had been catalogued in a century of mesmeric experiments—appeared to be contrived through unconscious expectation.

In the early decades of the twentieth century, hypnotic research in Russia and communist Europe followed a broadly Pavlovian framework, arguing that hypnotic suggestion, whether induced through physical passes or verbal instruction, worked on established physiological principles of inhibition. For Pavlov, hypnotically induced inhibition directed the cortex into "a partial, fragmentary, strictly localised sleep."[2]

Meanwhile, in North America, hypnosis found a host of new clinical and forensic adherents, yet it was only in the aftermath of World War II that experimental work on hypnotic induction, susceptibility, and the psychology of trance resumed in earnest. At the forefront of this new vanguard was the Stanford University Laboratory of Hypnosis Research, which, under the directorship of Ernest Hilgard, forged new measures of susceptibility and isolated the signature features of trance: mental inertia and planning avoidance; selective attention; heightened suggestibility; visual recall and fantasy-production; reduced reality testing; role-playing; and amnesia for the hypnotic experience. According to the Stanford Hypnotic Susceptibility Scale, fewer than five percent of the general population were found to be highly susceptible to hypnotic induction, designating them "hypnotic virtuosos."[3]

While Hilgard's research evolved, seeking to understand trance phenomena in terms of dissociation and the creation of subordinate cognitive systems, most of his findings were disputed by social psychologists who placed explanatory emphasis on role-enactment and the peculiar "demand characteristics" of the hypnotic experiment. Outside the laboratory, a powerful and enduring folklore of trance was also beginning to take root. Alongside media reports of "highway hypnosis" and "hypno-crime," sensational cases of recovered memory and past life regression, and a growing literature on brain-washing and mind-control, the dangers and possibilities of mental suggestion found a welcome home in a succession of feverish conspiracy theories and moral panics.

Three centuries after Mesmer, modern-day psychology remains enthralled with the same mystery: how exactly do the rituals of trance induction, the mental gymnastics of hypnotic rapport, allow control of mind and body to be ceded to a third party?

*

Animal magnetism introduced itself to pre-revolutionary Paris via a deluge of handbills. The tub, or baquet, described in the below advertisement, allowed Mesmer to scale his medical operation, but it introduced an unforeseen and unacknowledged element to magnetic treatment—social contagion.

ROBERT DARNTON, *MESMERISM AND THE END OF THE ENLIGHTENMENT IN FRANCE* (1968)

M. Mesmer, doctor of medicine of the faculty of Vienna in Austria, is the sole inventor of animal magnetism. That method of curing a multitude of ills (among others, dropsy, paralysis, gout, scurvy, blindness, accidental deafness) consists in the application of a fluid or agent that M. Mesner directs, at times with one of his fingers, at times with an iron rod that another applies at will, on those who have recourse to him. He also uses a tub, to which are attached ropes that the sick tie around themselves, and iron rods, which they place near the pit of the stomach, the liver, or the spleen, and in general near the part of their bodies that is diseased. The sick, especially women, experience convulsions or crises that bring about their cure. The mesmerizers (they are those to whom Mesmer has revealed his secret, and they number more than one hundred, including some of the foremost nobles of the court) apply their hands to the sick part and rub it for a while. That operation hastens the effect of the ropes and the rods. There is a tub for the poor every other day. In the antechamber, musicians play tunes likely to make the sick cheerful. Arriving at the home of this famous doctor, one sees a crowd of men and women of every age and state, the cordon bleu, the artisan, the doctor, the surgeon. It is a spectacle worthy of sensitive souls to· see men distinguished by their birth and their position in society mesmerize with gentle solicitude children, old people, and especially the indigent.[4]

Many of Mesmer's most active followers and radical champions belonged to the Society of Universal Harmony, which disseminated their master's "secret" through introductory courses and lectures, and a stream of evangelizing pamphlets. Among the early members of the Paris Society was the Marquis de Puységur (1751–1825) whose trials at his Buzancy estate were destined to reveal hidden depths of the liminal state that he called artificial somnambulism.

GEORGE MAKARI, *SOUL MACHINE: THE INVENTION OF THE MODERN MIND* (2015)

In 1784, the peak year of the Mesmer craze, the marquis began to take the health of the peasants on his estate into his own hands. One of his first cases was a stableboy, the twenty-three-year-old Victor Race, who was in bed for four days with pains and fever. Victor suffered from lung congestion, for which the marquis employed Mesmer's technique. To Puységur's astonishment, the young man fell into a waking sleep—"magnetic sleep," the marquis would later call it. The normally taciturn, obedient peasant, though clearly not awake, spoke candidly of his private life and its miseries. The marquis instinctively took it upon himself to combat these negative thoughts with more joyful ones: Victor had won a prize! He was dancing at a fair! He was singing! When Race awoke, he remembered nothing. Soon, his health returned.

What had transpired? . . . For Mesmer, the magnetizer exerted his will as a conductor of a force that linked two nervous systems into one. For a short while, their animal machines were wired together. However, the Marquis de Puységur was confronted not with two magnets in movement, but two subjects controlled by one will, his own. Victor absorbed the marquis's ideas without any hint of his own likes and dislikes. He had surrendered his ability to choose, want, or freely act. In magnetic sleep, his mind lost one of its most distinguishing features: intentional volition. In this weird state, the mind did not just become will-less but, according to the marquis, it became clairvoyant. Magnetized humans could gaze into the past and future, as well as peer through clothing and inside bodies.

To make matters even more astonishing, Puységur found that he did not even need to speak to convey his thoughts to Victor. Thoughts simply transferred from one mind to the other. He could arrest Victor's associations, change and rearrange them: he had complete control over the peasant's inner life. As lord of the estate, he already exercised a great deal of command over his peasants, but now the marquis could tell them how and what to think. He paused, however, when he considered the danger of this power falling into the wrong hands. An evil magnetizer could penetrate the secrets of others, abuse their confidence, and wreak havoc.

Magnetic sleep, he concluded, was some previously unknown form of somnambulism, one that could now be artificially invoked. With this, the marquis discovered another Victor inside Victor Race, and perhaps another self inside all selves.[5]

As animal magnetism travelled through Europe and the colonies, it discovered new audiences, effects, and uses. In the French colony of Saint-Domingue (Haiti), prior to the Haitian revolution, elements of magnetism found a place in voodoo healing and religion.

FRANCOIS REGOURD, "MESMERISM IN SAINT-DOMINGUE" (2008)

A magnetizer has been in the colony for a while now, and, following Mesmer's enlightened ideas, he causes in us effects that one feels without understanding them. We faint, we suffocate, we enter into truly dangerous frenzies that cause onlookers to worry. At the second trial of the tub, a young lady, after having torn nearly all her clothes, amorously attacked a young man on the scene. The two were so deeply intertwined that we despaired of detaching them, and she could be torn from his arms only after another dose of magnetism. You'll admit that such are ominous effects to which women should sooner not expose themselves. It produces a conflagration that consumes us, an excess of life that leads us to delirium. We will soon see a maltreated lover using it to his advantage.[6]

Like many of his mystically minded contemporaries, Heinrich Jung-Stilling
(1740–1817), a German physician and professor of political economy, saw
loftier possibilities in the trance state induced by mesmerism. Magnetic
somnambulists were, Jung-Stilling believed, gnostic prodigies, attuned to both
inner forces and higher powers.

JOHANN HEINRICH JUNG-STILLING, *THEORY OF PNEUMATOLOGY* (1808)

According to our common conceptions of human nature, the fact is astonishing, incomprehensible, and most remarkable, that all somnambulists, even the most vulgar and uneducated people, begin clearly to recognise their bodily illness, and even prescribe the most appropriate medicines for themselves, which the physician must also make use of if he wishes to gain his end. Even if they do not know the names of the remedies, yet they describe their qualities so minutely that the physician can soon ascertain them. In this state, also, they speak high German, where this is the language of the pulpit and the written tongue.

It is also very remarkable that somnambulists, who have often been in this state and at length attain this clearness of vision, arise, perform all kinds of work, play on an instrument if they have been taught music, go out to walk, &c., without their bodily senses having even the smallest perception of the visible world: they are then in the state of common sleep-walk. Thus it happened, that while I was at Bremen, in the autumn of the year 1798, a young woman came to me to ask my advice respecting her eyes. She was a somnambulist, and had herself decided upon consulting me in the crisis; her mother accompanied her, but she awoke in my presence, and I was therefore obliged to prescribe the appropriate remedies alone and without her assistance.

All these incidents, and others still more wonderful, may be found in the writings of the abovementioned authors. The most eminent physicians, and, generally speaking, every learned and rational thinking person, who has had the opportunity and the will to examine, with precision, the effects of animal magnetism, will attest that all that has been now advanced is pure truth, and confirm it by their testimony. But how is it that no one has hitherto attempted to draw hence those fertile inferences, by which the knowledge of human nature might be so much increased? To the best of my knowledge, no one has yet done so. Truly, so long as materialism is considered the only true system, it is impossible to comprehend such wonderful things; but, according to my system of theocratic liberty, not only is the whole comprehensible, but we are also led by magnetism to the most important discoveries, which before were only mysterious enigmas.[7]

Sometime philosopher, preacher, and gambler, Abbé José Custódio de Faria (1756–1819), began offering courses in magnetism in Paris in 1811. Professing no secret, claiming no special power, Fabia succeeded in ushering thousands of subjects into "lucid sleep" through verbal suggestion alone.

ABBÉ FARIA, "ON THE CAUSE OF LUCID SLEEP" (1819)

First of all, I reject all theories of animal magnetism, baquets, external will, magnetic fluid as being unnatural and extravagant. Lucid sleep is a matter as natural as memory and imagination, faculties common to everyone but not to the same extent. So also everyone is not equally susceptible to lucid sleep. Lucid sleep can be developed through intellectual practice and favourable physical conditions. But it has nothing to do with the external will of the concentrator since subjects can be made to fall into lucid sleep with will, without will or even with unexpressed opposite will. . . .

Suggestion, which is an order from a concentrator, is the immediate cause that triggers the real and precise cause that produces a particular and natural effect, but cannot produce it on its own. Induced lucid sleep, is a concentration of the senses produced at will and limited only by internal freedom, but caused by the external influence of the concentrator's suggestion.

The effects of lucid sleep are as ancient as the cradle of humanity, but have been noticed only recently in Europe by philosophical observers. Subjects with aptitude for lucid sleep are a challenge to human reason because they have deep knowledge on a variety of subjects, which they have acquired without study or meditation. They control all their involuntary movements; they reach objects at any distance in time or place and, consequently, through all obstacles; they read without help from the eyes any book, even closed; they unveil thoughts, even constant ones;

they cause thousands of other sensory and real effects. But hoping to find in these oracles' predictions an unclouded truth, is a delusion and vain expectation that will never be fulfilled.[8]

Faria's misgivings regarding the oracular pronouncements of magnetized subjects went unheeded in occult and Spiritualist circles. Louise Alphonse Cahagnet's Celestial Telegraph *described experiments with eight "ecstatic somnambulists" which furnished evidence of the "life to come."*

LOUISE ALPHONSE CAHAGNET, *THE CELESTIAL TELEGRAPH* (1851)

At one sitting, she tells me that she can see her father and mother, who had been dead a long time, that they appear to her really alive. Her mother was surrounded by a beautiful blue sky, on her right, and ventured not to approach her; she was precisely in the same dress as she wore before her death; her father, too, was attired as on earth, and sought to conceal himself behind a bush on her left. She contemplated them a long time without speaking to them. This was the first day she had been clairvoyant, and perceived her hopeless condition, assuring me that if I continued to magnetize her I should have to take the greatest precautions, for my life was at stake, as I had been already told by two clairvoyants. I took no further account of these warnings than by guarding against her baneful emanations. On awaking, she recollected having seen her father and mother, but had not the least recollection of her hopeless condition, or what she had said respecting it.[9]

Among the medical champions of mesmerism, there was an almost messianic belief in its regenerative possibilities. The prominent London physician John Elliotson founded The Zoist *in the belief that the science of mesmerism could bring about "the progressive improvement and increasing happiness of the race."*

JOHN ELLIOTSON, PROSPECTUS FOR *THE ZOIST* (1843)

The discovery of a new truth gives to the philosopher intense delight. The science of MESMERISM is a new physiological truth of *incalculable* value

and importance; and though sneered at by the pseudo-philosophers of
the day, there is not the less certainty that it presents the only avenue
through which is discernible a ray of hope that the more intricate phenomena of the nervous system,—of Life,—will ever be revealed to man.
Already has it established its claim to be a most potent remedy in the cure
of disease; already enabled the knife of the operator to traverse and divide
the living fibre unfelt by the patient. If such are the results of its infancy,
what may not its maturity bring forth? . . . Let what many facts render
probable be once established, viz., that this state of increased activity can
be rendered permanent and carried into the natural state, and who does
not catch a glance of a mighty engine for man's regeneration, vast in its
power and unlimited in its application, rivalling in morals the effects of
steam in mechanics.[10]

*First-person descriptions of the trace state are surprisingly rare in the medical
and philosophical literature on animal magnetism. In the 1840s, the
prominent writer and long-suffering invalid Harriet Martineau (1802–1876)
sought out the services of a mesmerist, Spencer Hall, who visited her bedside
and provided temporary relief from a uterine disorder.*

HARRIET MARTINEAU, *LETTERS ON MESMERISM* (1845)

Within one minute the twilight and phosphoric lights appeared; and in
two or three more, a delicious sensation of ease spread through me,—a
cool comfort, before which all pain and distress gave way, oozing out, as
it were, at the soles of my feet. During that hour, and almost the whole
evening. I could no more help exclaiming with pleasure than a person
in torture crying out with pain. I became hungry, and ate with relish,
for the first time for five years. There was no heat, oppression, or sick-
ness during the seance, nor any disorder afterwards. During the whole
evening, instead of the lazy hot ease of opiates, under which pain is felt
to lie in wait, I experienced something of the indescribable sensation of
health, which I had quite lost and forgotten. I walked about my rooms,
and was gay and talkative. Something of this relief remained till the next
morning; and then there was no reaction. I was no worse than usual; and
perhaps rather better.

Nothing is to me more unquestionable and more striking about this influence than the absence of all reaction. Its highest exhilaration is followed, not by depression or exhaustion, but by a further renovation. From the first hour to the present, I have never fallen back a single step. Every point gained has been steadily held. Improved composure of nerve and spirits has followed every mesmeric exhilaration. I have been spared all the weaknesses of convalescence, and have been carried through all the usually formidable enterprises of return from deep disease to health with a steadiness and tranquility astonishing to all witnesses. At this time, before venturing to speak of my health as established, I believe myself more firm in nerve, more calm and steady in mind and spirits, than at any time of my life before. So much, in consideration of the natural and common fear of the mesmeric influence as pernicious excitement—as a kind of intoxication.[11]

Public knowledge of mesmerism across Europe in the early decades of the nineteenth century relied mostly on lectures and demonstrations by itinerant "professors." One such performer, Charles Lafontaine (1803–1892) began lecturing in Britain in 1841, attempting to "induce doctors and men of science to inquire into [its agency]".

CHARLES LAFONTAINE, *MEMOIRS OF A MAGNETISER* (1866)

Having accomplished the cure of numerous deaf and blind persons, as also of numerous epileptic and paralytic sufferers, at the Hospital [in Birmingham], I repaired to Liverpool, but only to meet with disappointment; few persons attended the séance; and on the following day I proceeded to Manchester, one of those cities in which my success was conspicuous. The newspapers reported my experiments at great length, filling four or five columns of the colossal English journals, and to give some idea of the sensation which I created in this great manufacturing centre, I may say that my séances returned me a gross total of thirty thousand francs, though the charge for admission was only half-a-crown. The concourse was truly immense, and the lecture hall of the Athenaeum was more than crowded. I put to sleep a number of persons who were well-known residents of Manchester, occupying good positions, among

others Messrs Lynnil, Higgins, Dyrenfurth, a gentleman on the staff of
the Guardian, and several more. I caused deaf mutes to hear, operated a
number of different brilliant cures, and then retired to Birmingham a second time, where I had an engagement to visit a patient, and was to give
some further demonstrations.[12]

Following Lafontaine's demonstrations of mesmerism in Manchester, the
Scottish surgeon James Braid (1795–1860) undertook a series of trials that led
him to propose an alternative nomenclature.

JAMES BRAID, "LETTER TO *THE LANCET*" (1845)

I adopted the term "hypnotism" to prevent my being confounded with
those who entertain those extreme notions [sc. that a mesmeriser's *will*
has an "irresistible power . . . over his subjects" and that clairvoyance and
other "higher phenomena" are routinely manifested by those in the mesmeric state], as well as to get rid of the erroneous theory about a magnetic
fluid, or exoteric influence of any description being the cause of the sleep.
I distinctly avowed that hypnotism laid no claim to produce any phenomena which were not "quite reconcilable with well-established physiological and psychological principles"; pointed out the various sources
of fallacy which might have misled the mesmerists; [and] was the first to
give a public explanation of the trick [by which a fraudulent subject had
been able to deceive his mesmerizer]. . . .

[Further, I have never been] a supporter of the imagination theory—
i.e., that the induction of [hypnosis] in the first instance is merely the
result of imagination. My belief is quite the contrary. I attribute it to the
induction of a habit of intense abstraction, or concentration of attention,
and maintain that it is most readily induced by causing the patient to fix
his thoughts and sight on an object, and suppress his respiration.[13]

Though operations under mesmeric anesthesia were reported as early as the
1820s, the medical establishment remained deeply suspicious of mesmerism's
surgical efficacy. James Esdaile (1808–1859), a physician employed by the
East India Company, found otherwise, pioneering its use with native patients
at Calcutta's Hooghly Imambara Hospital.

JAMES ESDAILE, *MESMERISM IN INDIA* (1846)

A Return showing the Number of painless Surgical Operations performed at Hooghly, during the last eight months.

Arms amputated	1
Breast ditto	1
Tumour extracted from the upper jaw	1
Scirrhus testium extirpated	2
Penis amputated	2
Contracted knees straightened	3
Ditto arms	3
Operations for cataract	3
Large tumour in the groin cut off	1
Operations for Hydrocele	7
Ditto Dropsy	2
Actual Cautery applied to a sore	1
Muriatic acid ditto	2
Unhealthy sores pared down	7
Abscesses opened	5
Sinus, six inches long, laid open	1
Heel flayed	1
End of thumb cut off	1
Teeth extracted	3
Gum cut away	1
Prepuce cut off	3
Piles ditto	1
Great toenails cut out by the roots	5
Seton introduced from ankle to knee	1
Large tumour on leg removed	1
Scrotal tumours, weighing from 8 lb to 80lb removed 17	<u>14</u>
Painless operations	<u>73</u>

I beg to state, for the satisfaction of those who have not yet a practical knowledge of the subject, that I have seen no bad consequences whatever arise from persons being operated on when in the mesmeric trance. Cases have occurred in which no pain has been felt subsequent to the operation even; the wounds healing in a few days by the first intention; and in the rest, I have seen no indications of any injury being done to the constitution. On the contrary, it appears to me to have been saved, and that less constitutional disturbance has followed than under ordinary circumstances.

There has not been a death among the cases operated on. In my early operations, I availed myself on the first fit of insensibility, not knowing whether I could command it back again at pleasure.

But if the trance is not profound the first time, the surgeon may safely calculate on its being deeper the next, and when operating in public, it will be prudent to take the security of one or two preliminary trances. Flexibility of the limbs till moved, and their remaining rigid in any position we put them in, are characteristic of a trance: but there are exceptions, and these are equally diagnostic, and to be depended upon. It sometimes happens that the limbs become rigid as they lie, and on bending them they have always a disposition to return to a state of spasmodic extension. At other times, there is a complete relaxation of the whole muscular system, and the limbs can be tossed about like those of a person just dead.[14]

Jean-Martin Charcot, the so-called "Napoleon of Neuroses," was the most famous physician in Europe, when, in 1878, he began to investigate the effects of hypnotism on hysterical patients on the wards of Salpêtrière hospital. Swedish-born physician and psychiatrist Axel Munthe, a student of Charcot in the 1880s, echoed misgivings that were already widespread when he found the Salpêtrière's somnambulists to be participants in a semi-scripted, three-act drama.

AXEL MUNTHE, *THE STORY OF SAN MICHELE* (1929)

To me who for years had been devoting my spare time to study hypnotism these stage performances of the Salpêtrière before the public of Tout Paris were nothing but an absurd farce, a hopeless muddle of truth and

cheating. Some of these subjects were no doubt real somnambulists faith-
fully carrying out in a waking state the various suggestions made to them
during sleep—posthypnotic suggestions. Many of them were mere frauds,
knowing quite well what they were expected to do, delighted to perform
their various tricks in public, cheating both doctors and audience with the
amazing cunning of the hysteriques. They were always ready to 'piquer
une attaque' of Charcot's classical grande hystérie, arc-en-ciel and all, or
to exhibit his famous three stages of hypnotism: lethargy, catalepsy, somnambulism, all invented by the Master and hardly ever observed outside
the Salpêtrière. Some of them smelt with delight a bottle of ammonia
when told it was rose water, others would eat a piece of charcoal when
presented to them as chocolate. Another would crawl on all fours on the
floor, barking furiously, when told she was a dog, flap her arms as if trying to fly when turned into a pigeon, lift her skirts with a shriek of terror
when a glove was thrown at her feet with a suggestion of being a snake.
Another would walk with a top hat in her arms rocking it to and fro and
kissing it tenderly when she was told it was her baby. Hypnotized right
and left, dozens of times a day, by doctors and students, many of these
unfortunate girls spent their days in a state of semi-trance, their brains
bewildered by all sorts of absurd suggestions, half conscious and certainly
not responsible for their doings, sooner or later doomed to end their days
in the *salle des agités* if not in a lunatic asylum.[15]

*With the demise of Charcot's physiological, three-stage theory of hypnosis, the
Nancy School, led by Ambroise-Auguste Liébeault and Hippolyte Bernheim
prevailed. Positing suggestion as the key to understanding hypnotism,
Bernheim (1840–1919) insisted that all its effects were fashioned from the
(ill-defined) alchemy of expectation between operator and subject.*

HIPPOLYTE BERNHEIM, *SUGGESTIVE THERAPEUTICS* (1880)

No: hypnotic sleep is not a pathological sleep. The hypnotic condition
is not a neurosis, analogous to hysteria. No doubt, manifestations of
hysteria may be created in hypnotized subjects; a real hypnotic neurosis
may be developed which will be repeated each time sleep is induced.
But these manifestations are not due to the hypnosis,—they are due to

the operator's suggestion, of sometimes to the auto-suggestion of a particularly impressible subject whose imagination, impregnated with the ruling idea of magnetism, creates these functional disorders which can always be restrained by a quieting suggestion. The pretended physical phenomena of hypnosis are only psychical phenomena. Catalepsy, transfer, contracture, etc., are the effects of suggestion. To prove that the very great majority of subjects are susceptible to suggestion is to eliminate the idea of a neurosis. At least it is not an admission that the neurosis is universal, that the word hysteria is a synonym for any nervous impressionability whatever. For, as we all have nervous tissues, and as it is a property of such tissues to be impressionable, we should all be hysterical. The sleep itself is the effect of a suggestion. I have said that no one could be hypnotized against his will. M. Ochorowitz has opposed this proposition energetically. Perhaps he has not quite grasped my idea. It is certain that any one who does not want to be hypnotized, and who knows that he need not be influenced if he does not wish to be, successfully resists every trial. It is also true that certain subjects cannot resist because their will-power is weakened by fear, or by the idea of a superior power which influences them in spite of themselves. No one can be hypnotized unless he has the idea that he is going to be. Looked at in this light my proposition cannot be attacked. The idea makes the hypnosis; it is a psychical and not a physical or fluid influence which brings about this condition. It is a singular thing that psychologists like M. Janet and M. Binet have failed to recognize the purely psychical nature of these manifestations. M. Delboeuf has not been deceived in the matter.[16]

Though Sigmund Freud had personally observed Bernheim's "astonishing experiments," gaining "the profoundest impression of the possibility that there could be powerful mental processes which nevertheless remained hidden from the consciousness of man," he had limited success with hypnosis in his own practice.

SIGMUND FREUD, LETTER TO WILHELM FLIESS (MAY 28, 1888)

I have at this moment a lady in hypnosis lying in front of me and therefore can go on writing in peace.[17]

SIGMUND FREUD, *GROUP PSYCHOLOGY AND THE ANALYSIS OF EGO* (1921)

From being in love to hypnosis is evidently only a short step. The respects in which the two agree are obvious. There is the same humble subjection, the same compliance, the same absence of criticism, towards the hypnotist as towards the loved object. There is the same sapping of the subject's own initiative; no one can doubt that the hypnotist has stepped into the place of the ego ideal.[18]

According to early commentators, "artificial somnambulists" were generally impervious to improper or questionable overtures. It was not until the 1860s that practitioners and experimenters began to investigate the limits of hypnotic compliance.

ALBERT MOLL, *HYPNOTISM* (1890)

There is no doubt that subjects may be induced to commit all sorts of imaginary crimes in one's study. I have made hardly any such suggestions, and have small experience on the point. In any case a repetition of them is superfluous. If the conditions of the experiment are not changed, it is useless to repeat it merely to confirm what we already know. And these criminal suggestions are not altogether pleasant. I certainly do not believe that they injure the moral state of the subject, for the suggestion may be negatived and forgotten. But these laboratory experiments prove nothing, because some trace of consciousness always remains to tell the subject he is playing a comedy.[19]

Away from the hospital wards and clinics, the comedic aspects of the trance performance remained central to its popular appeal.

JAMES COATES, *HOW TO MESMERISE* (1893)

Lead all your subjects into a garden (an imaginary one). Let them behold its beauties, enjoy its fruits, and gather its flowers. Give them full freedom of action and speech therein. . . . Discover a cluster of bees swarming in

some corner of the garden. Let someone, thoughtlessly or greedily searching for fruit, disturb the bees. The change of scene is magical. Some will
desperately fight the bees; others will manifest rage; some will sit down
and try to cover themselves, while others will cry like children.[20]

For the poet, classicist, and psychical researcher Frederic Myers (1843–1901),
even the most prosaic demonstrations of mesmeric suggestion exposed a
doorway into the "wider mystery" of the subliminal self.

F. W. H. MYERS, *HUMAN PERSONALITY* (1909)

Self-suggestion, whatever this may really mean, is thus in most cases [of
hypnosis], whether avowedly or not, at the bottom of the effect produced. It has already been used most successfully, and it will probably
become much commoner than it now is;—or, I should rather say (since
every one no doubt suggests to himself when he is in pain that he would
like the pain to cease), I anticipate that self-suggestion, by being in some
way better directed, will become more *effective*, and that the average of
voluntary power over the organism will rise to a far higher level than it
at present reaches. I believe that this is taking place even now; and that
certain *schemes of self-suggestion*, so to call them, are coming into vogue,
where patients in large masses are supplied with effective conceptions,
which they thus impress repeatedly upon themselves without the need
of a hypnotiser's attendance on each occasion. The "Miracles of Lourdes"
and the cures effected by "Christian Science" fall, in my view, under this
category. We have here suggestions given to a quantity of more or less
suitable people *en masse*, much as a platform hypnotiser gives suggestions to a mixed audience, some of whom may then be affected without
individual attention from himself. . . .

I do not indeed pretend that my explanation can enable us to reduce
hypnotic success to a certainty. I cannot say why the process should be so
irregular and capricious; but I can show that this puzzle is part and parcel of a wider mystery;—of the obscure relationships and interdependen-
cies of the supraliminal and the subliminal self. In split personalities, in
genius, in dreams, in sensory and motor automatisms, we find the same
fitfulness, the same apparent caprice.[21]

With the publication of Emile Coué's (1857–1926) best-selling 1922 book, self-suggestion became a widely practiced method for alleviating pain and distress on both sides of the Atlantic.

EMILE COUÉ, *SELF MASTERY THROUGH CONSCIOUS AUTOSUGGESTION* (1922)

Before sending away your patient, you must tell him that he carries within him the instrument by which he can cure himself, and that you are, as it were, only a professor teaching him to use this instrument, and that he must help you in your task. Thus, every morning before rising, and every night on getting into bed, he must shut his eyes and in thought transport himself into your presence, and then repeat twenty times consecutively in a monotonous voice, counting by means of a string with twenty knots in it, this little phrase: "EVERY DAY, IN EVERY RESPECT, I AM GETTING BETTER AND BETTER." In his mind he should emphasize the words "in every respect" which applies to every need, mental or physical. This general suggestion is more efficacious than special ones.[22]

Colgate University psychologist George Hoben Estabrooks (1895–1973) claimed to have conducted military field trials to create hypnotized assassins and programmed spies. Shunned by the intelligence community, Estabrooks helped fuel the growing mythology of mind control, claiming that he could "hypnotize a man—without his knowledge or consent—into committing treason against the United States."

GEORGE HOBEN ESTABROOKS, *HYPNOTISM* (1943)

The only possible way of determining whether or not a subject will commit a murder in hypnotism is literally to have him commit one. No "fake" setup will satisfy the critics, for the hypnotised subject is not "asleep." He is very wide awake, willing to co-operate in all kinds of fake murders with rubber knives. But with a real knife or a loaded revolver? No one knows, for the simple reason that no one dares find out. The police would not see the point when they viewed the corpse and were told it was the result of a "scientific" experiment. Nor would the jury. Sing Sing and the electric chair would probably put an end to the career of the particular "scientist."

But warfare may, undoubtedly will, answer many of these questions. A nation fighting with its back to the wall is not very worried over the niceties of ethics. If hypnotism can be used to advantage we may rest assured that it will be so employed. Any "accidents" which may occur during the experiments will simply be charged to profit and loss, a very trifling portion of that enormous wastage in human life which is part and parcel of war.[23]

Shortly after the publication of George Orwell's Nineteen Eighty-Four, Aldous Huxley wrote to Orwell, his former student at Eton, contemplating the use of hypnotism in the "ultimate revolution."

ALDOUS HUXLEY, LETTER TO GEORGE ORWELL (1949)

Within the next generation I believe that the world's rulers will discover that infant conditioning and narco-hypnosis are more efficient, as instruments of government, than clubs and prisons, and that the lust for power can be just as completely satisfied by suggesting people into loving their servitude as by flogging and kicking them into obedience. In other words, I feel that the nightmare of *Nineteen Eighty-Four* is destined to modulate into the nightmare of a world having more resemblance to that which I imagined in *Brave New World*. The change will be brought about as a result of a felt need for increased efficiency. Meanwhile, of course, there may be a large-scale biological and atomic war—in which case we shall have nightmares of other and scarcely imaginable kinds.[24]

Two dominant theories of the trance state emerged in post-war American psychology: the social-psychological and the neodissociationist. The social-psychological model found that the hypnotic subject was responding to social cues and demand characteristics embedded in the experimental setting.

THEODORE R. SARBIN, "CONTRIBUTIONS TO ROLE-TAKING THEORY" (1950)

It appears the stage director stands in the same relationship to the actor as the hypnotist does to the subject. The statuses of positions are defined beforehand, the specific role-behaviours are dictated by the attempts of each participant to validate his status. In short, the participants inter-behave with each other in ways that are appropriate to each position—provided, of course, that such interbehavior can be incorporated by each participant in his self-concept. Because acting has not been burdened with the incubus of dissociation or ideomotor theory, we are not amazed at the frequent marked changes in skeletal and visceral behavior which occur merely because the director tells the actor what to do. The analyst of dramatic acting does not seem to be concerned with such pseudo-problems as the search for a one-to-one constancy relationship between the magnitude of the stimulus (the director's verbal instructions) and the magnitude of the response (the complicated verbal, motor, and visceral reactions of the actor).[25]

With the advent of the Cold War, journalist and propaganda expert Edward Hunter began to seed the idea that the spread of Communism was aided by shadowy methods of hypnotic indoctrination, which he dubbed brain-washing.

EDWARD HUNTER, *BRAIN-WASHING IN RED CHINA* (1951)

Brain-washing became the principal activity on the Chinese mainland when the Communists took over. Unrevealed tens of thousands of men, women, and children had their brains washed. They ranged from students to instructors and professors, from army officers and municipal officials to reporters and printers, and from criminals to church deacons. There were no exceptions as to profession or creed. Before anyone could be considered trustworthy, he was subjected to brain-washing in order to qualify for a job in the "new democracy." Only then did the authorities consider that he could be depended upon, as the official expression is worded, to "lean to one side" (Soviet Russia's) in all matters, and that he would react with instinctive obedience to every call made upon him by the Communist Party through whatever twists, turns, or leaps policy might take, no matter what the sacrifice. He must fight by all possible means and be ready, too, with the right answer for every contradiction and evasion in Party statements.[26]

Thanks to the efforts of Michigan-born market researcher James Vicary (1915–1977), the equally dubious notion of subliminal advertising began to gain commercial and public credence.

"SUBLIMINAL PROJECTION," *ADVERTISING AGE* (1957)

Market researcher James M. Vicary today unveiled his secret new weapon for advertisers—the 'invisible commercial.' It is based on the theory of 'subliminal' projection. Assuming the idea is feasible this will enable advertisers to flash sales messages on tv without the viewers being consciously aware of them. The messages will reach the audience subliminally—that is, 'below the threshold of sensation or awareness.'

Mr. Vicary showed reporters a film interlarded with Coca-Cola commercials. The Coke messages were flashed at the rate of one every five seconds, and only a few of them were detected by the audience. Mr. Vicary explained that these few were visible because he rigged the mechanism so that the reporters could see what was being done visibly.

Mr. Vicary, head of the motivation research company bearing his name, said the commercial messages are superimposed on a film as 'very brief overlays of light.' They are so rapid—up to 1/3000 of a second—that they cannot be seen by the audience.

Mr. Vicary reported that he recently tested the 'invisible commercial' in a New Jersey movie theater. The tests ran for six weeks, during which time some 45,000 persons attended the theater. Two advertising messages were projected—one urging the audience to eat popcorn, the other suggesting, 'Drink Coca-Cola.'

According to Mr. Vicary, the 'invisible commercial' increased popcorn sales by 57.5% and Coca-Cola sales by 18.1%.

Mr. Vicary emphasized that his subliminal ads are useful only as reminder advertising. They will not, he said, move a person to switch brands.[27]

From Ray Bradbury's Fahrenheit 451 *to Anthony Burgess'* A Clockwork Orange, *post-war popular fiction routinely explored the authoritarian and commercial possibilities of psychological conditioning. Richard Condon's pulp classic* The Manchurian Candidate *leaned a little too heavily on sensationalist exposes of Communist "brainwashing."*

RICHARD CONDON, *THE MANCHURIAN CANDIDATE* (1959)

"Brainwashing?" Raymond did not like that note. He could not abide the thought of anybody tampering with his person so he rejected the entire business then and there. . . . The disgust it made Raymond feel acted like a boathook that pushed the solid shore away from him to allow him to drift away from it on the strong-flowing current of self. It did not mean that he had instantly closed his mind to Marco's problem. He most earnestly wanted to be able to help Ben find relief, to help to change his friend's broken mechanism, to find him sleep and rest and health, but his own participation in what he had started out to make a flaming patriotic crusade when he had first started to speak had been muted by his fastidiousness: he shrank from what he could only consider the rancid vulgarity of brainwashing.

"It has to be a brainwash," Marco said intensely. "In my case it slipped. In Melvin's case it slipped. It's the only possible explanation, Raymond. The only, only explanation."

"Why?" Raymond answered coldly. "Why would the Communists want me to get a Medal of Honor?"

"I don't know. But we have to find out."[28]

The supernatural aspects of the trance state, first described by the somnambules of the early nineteenth century, returned to the fore with the publication of The Search for Bridey Murphy, *a sensational tale of reincarnation.*

MARTIN GARDNER, "BRIDEY MURPHY AND OTHER MATTERS" (1957)

Under hypnosis, a brown-haired, trim-figured little housewife, Mrs. Virginia Tighe, began to talk in Irish brogue about her previous incarnation as a red-headed colleen named Bridey Murphy. William J. Barker, assistant editor of the *Denver Post's* Sunday supplement, *Empire*, serialized the story in 1954 (Sept. 12, 19, and 26) under the title, "The Strange Search for Bridey Murphy." The reader response suggested that here was material for a national best-seller. Morey Bernstein, the Pueblo businessman who had hypnotized Virginia, decided to write a book about it. Barker helped on the manuscript and Doubleday printed it in 1956 as *The Search for Bridey Murphy*.

For many weeks the book topped the country's best-seller lists. It was translated into five other languages. A tape recording of one of Mrs. Tighe's trance sessions was placed on a long-playing record and tens of thousands of copies were sold at $5.95 each. *True* magazine condensed the book. More than forty newspapers syndicated it. Movie rights were sold. Hostesses gave "Come as you were" parties. Juke boxes blared *Do You Believe in Reincarnation?*, *The Love of Bridey Murphy*, and the *Bridey Murphy Rock and Roll*. Night club hypnotists who hadn't worked for years suddenly found themselves in great demand. All over the country, and especially in California, amateur hypnotists began sending parlor subjects back to previous lives. One lady described her existence in 1800 as a horse. In Shawnee, Oklahoma, a teen-age boy shot himself, leaving a note saying that he was curious about the Bridey theory and would now investigate it in person. Two studies of Edgar Cayce were rushed back into print simply because Bernstein mentioned them favorably. A rash of new books on hypnotism, reincarnation, and related occult topics broke out on publishers' lists. In the words of a Houston book dealer, Bridey was 'the hottest thing since Norman Vincent Peale.'

One would be hard put to find a choicer sample of an utterly worthless book designed to exploit a mass hunger for scientific evidence of life after death, or a better example of the power of modern huckstering to swindle the gullible, simple folk who take such books seriously. . . . If Virginia had been in the hands of a trained, well-informed psychologist, what would his reaction have been? In the first place he would have immediately recalled the classical cases exactly like Virginia's. Almost any hypnotic subject capable of going into a deep trance will babble about a previous incarnation if the hypnotist asks him to. He will babble just as freely about his future incarnations. Usually what he has to say is dreary and uninspired. At times, however, he spins such a detailed story that it becomes a matter of special interest.[29]

As hypnosis established itself as an adjunct to psychiatry and clinical psychology, patents were filed for devices such as the Brain Wave Synchronizer, a machine that claimed to accelerate hypnotic induction and receptivity.

WILLIAM S. KROGER AND SIDNEY A. SCHNEIDER, *AN ELECTRONIC AID FOR HYPNOTIC INDUCTION* (1959)

In operation, the patient assumes a comfortable position, sitting or reclining, facing the general direction of the instrument. The distance between patient and machine is not important. Instructions are given. The instrument is turned on and set to the approximate brain frequency. The synchronizer starts instantly—no warm-up time is required. After the subject is hypnotized, suggestions can be given with the instrument on or off. If left on, suggestions can be given to maintain the hypnotic state. The unit is extremely portable, weighing only five pounds.

As the clinical applications for hypnosis expand, so will the demand for electronic aids. The present application of the BWS [Brain Wave Synchronizer] is the beginning of a long range program which may have far reaching significance in subliminal projection and other fields of communication. As the functions of the reticular activating system and the alpha waves in their relationships to alterations in perception and consciousness are elucidated, more light will be shed on the worth and development of hypnosis-inducing devices.

The BWS has already proved its worth in medicine, dentistry and obstetrics. Further research is necessary to prove the effectiveness of this device in all fields to produce "dissociative" states of greater receptivity to suggestions. The authors believe that the instrument will have a salutary effect in the field of clinical and experimental hypnosis.[30]

The CIA handbook—using the cryptonym of KUBARK—urges caution in the field.

"HYPNOTISM AND THE CIA OPERATIVE," KUBARK COUNTERINTELLIGENCE INTERROGATION (1963)

Operational personnel, including interrogators, who chance to have some lay experience or skill in hypnotism should not themselves use hypnotic techniques for interrogation or other operational purposes. There are two reasons for this position. The first is that hypnotism used as an operational tool by a practitioner who is not a psychologist, psychiatrist, or

M.D. can produce irreversible psychological damage. The lay practitioner does not know enough to use the technique safely. The second reason is that an unsuccessful attempt to hypnotize a subject for purposes of interrogation, or a successful attempt not adequately covered by post-hypnotic amnesia or other protection, can easily lead to lurid and embarrassing publicity or legal charges.[31]

Meanwhile, hypnotic passivity was employed as a fast-track learning aid, spawning a growing number of sleep learning societies and institutes across Europe and North America.

LYNN SCHROEDER AND SHEILA OSTRANDER, *PSYCHIC DISCOVERIES BEHIND THE IRON CURTAIN* (1970)

'It's not hypnosis or sleep learning. It's far more practical than that. The student is fully awake and in complete control of himself,' says Dr Lozanov. It's a kind of mind to mind contact between teacher and student, based on Yoga techniques, and Lozanov calls it, 'Suggestopedia.'

In a typical classroom at the Institute, twelve people—students, housewives, laborers, professional people, old and young—relax in reclining chairs that resemble airplane seats. The room looks more like a lounge than a classroom. The lighting is subdued to enhance the calming effect. The students are listening to music, gentle soothing music. They look as if they were at a concert, completely wrapped up in the harmony of sounds.

In actuality this is a French lesson. Against the background of Brahms or Beethoven, the voice of the teacher seems sometimes businesslike, as if ordering work to be done, sometimes soft and calming, then unexpectedly hard, commanding. Her voice repeats in a special rhythm, on a special scale of intonation, French words, idioms, and their translations. But the students aren't really listening. They've been warned *not* to pay attention, not to think about whether they hear the teacher. 'Relax. Don't think about anything.' Their conscious minds are to be totally occupied with the music.

The next day surprised students discover that even though they were sure they'd learned nothing, they remember and can easily read, write, and speak from 120 to 150 new words absorbed during the two hour session. In the same way the toughest part of the language course, the grammar rules, painlessly take root in the minds of music-lulled students. Within a month students with no prior knowledge of a language have accepted two to three thousand vocabulary words and have a good grasp of the grammar. Tests a year later show they still know the material learned in this incredibly effortless way.[32]

American Laboratory research into hypnosis dwindled in the era of behaviorism. Though Clark Hull's Hypnosis and Suggestibility *(1933) signaled the dawning of a new experimental approach, two decades passed before Ernest Hilgard pioneered a new wave of research at Stanford University. The apotheosis of Hilgard's work, the neo-dissociationist theory of an "inner observer," was dismissed by some critics as "a pure laboratory artifact."*

ERNEST HILGARD, *THE HIDDEN OBSERVER* (1977)

Divided consciousness is familiar in ordinary waking life; the division permits fantasy to continue even while the person is performing the obligations of the work life or satisfying the proprieties of social interactions and communications. Because he is able to pay sufficient attention to the obligations and proprieties, this fantasizing may go largely unnoticed. In hypnosis the fantasied world may become prominent, and, at least in some instances, the realities may be denied. The question arises whether the denied realities are completely obliterated, or whether instead they are concealed behind an amnesia-like mask that is possible with hypnosis. We found in some demonstrations and experiments within our laboratory that two kinds of information processing may go on at once within hypnosis; some aspects are available to the hypnotic consciousness within hypnosis as ordinarily studied; other aspects are available only when special techniques have elicited the concealed information. When these techniques are used, the additional information is reported as though it had been observed in the usual manner. Because the observing part was hitherto not in awareness, we have come to use the metaphor of a 'hidden observer' to characterise the cognitive system. . . .[33]

Let the subjects speak for themselves:

The hidden part doesn't deal with pain. It looks at what is, and doesn't
   judge it.
It is not a hypnotized part of the self. It knows all parts.
The hidden observer is watching, mature, logical, has more information.

*Perhaps the most controversial use of modern-day hypnosis has been its
deployment as a means for recovering memories of trauma and abuse. In the
early 1990s, as psychologists began to take stock of the false and confabulated
memories that clinicians were unwittingly co-authoring, the psychiatrist John
Mack declared himself convinced by cases of alien abduction, which he had
also uncovered via hypnosis.*

JOHN MACK, *ABDUCTION: HUMAN ENCOUNTERS WITH ALIENS* (1995)

A final word needs to be said about the use of hypnosis in Ed's case.
Before my first meeting with him, Ed had recalled a great deal about
his teenage abduction experience. But his conscious memory before the
regression tended to simplify the experience and, more significantly, to
gloss the narrative in ways that were more syntonic with the self-image
and desires of a young adolescent than what he recalled painfully during the hypnosis session. Many embarrassing details relating to pow-
erlessness and loss of control were not available to him except under
hypnosis. In particular, the happy outcome of pleasurable sexual intercourse with the cooperative, sexually active, female alien gave way to the
forced, quite humiliating, taking of a sperm sample as the being watched
approvingly. This second scenario, which is obviously more disturbing,
is far more typical of male abduction experiences and, therefore, more
believable.

   All this suggests that, at least in Ed's case, the information recalled
painstakingly under hypnosis is more reliable than the consciously
recalled story, which seems to have been unconsciously adjusted to be
compatible with Ed's wishes and self-esteem. There are other details
obtained during the hypnosis session relating to the transport to the
craft, the numbers of beings (the female leader and her "staff" rather than

the single alien "woman"), the two chambers (an O.R.-like room and the podlike briefing room) rather than the single pod, and the great amount of information transmitted by the alien female, which make the story obtained during the regression more believable, or at least more consistent, with other abduction accounts.[34]

While neuroimaging technology promised to capture the deep signature of hypnosis, its laboratory recording of the trance state brought as much heat as it did light to the new science of hypnosis.

JOHN F. KIHLSTROM, "NEURO-HYPNOTISM" (2013)

Historically, the most popular approach to understanding the neural substrates of hypnosis has been to examine EEG correlates of hypnotizability and changes in the EEG spectrum which occur when hypnosis is induced. . . . Many of these studies were "fishing expeditions", conducted in the hopes that they would yield interesting results, rather than tests of specific hypotheses about the nature or locus of electrocortical changes associated with hypnosis. Still, they were not always without some theoretical rationale, however weak. For example, in the late 1960s it was suggested that hypnotizability and hypnosis were associated with increased density of alpha activity in the EEG—a hypothesis which drew strength from early reports of increased alpha density in Zen and yoga meditation, as well as the meditation-like experiences once thought to be produced by EEG alpha biofeedback. . . .

Perhaps the most provocative EEG finding was by Macleod-Morgan and Lack (1982), who found that hypnotizable subjects showed greater task-specific hemispheric activation than did their insusceptible counterparts. That is, hypnotizable subjects appeared more likely to activate the left hemisphere when performing a task designed to selectively activate the left hemisphere, and the right hemisphere when performing a right-hemisphere task. Although subsequent attempts to replicate have yielded somewhat mixed results, Macleod-Morgan and Lack's findings led to the revised hypothesis that hypnotizable subjects possessed a flexible cognitive style which permitted them to shift easily between analytic (left hemisphere) and holistic (right hemisphere) modes of processing, as demanded by the task at hand: this flexibility is further enhanced by the induction of hypnosis. [Further research has suggested that] hypnotizable subjects are adept at tasks involving either analytic or holistic processing, and at tasks involving either sustained attention or disattention, especially when they are hypnotized. Put another way, the hypnotizable brain, even when hypnotized, is just like any other brain—only better.[35]

4

OTHER WAYS OF KNOWING: OF INSTINCTS AND INTUITIONS

To think without awareness; to act without agency; to make decisions and solve problems without knowing deliberation—whether approached via the concept of intuition or instinct, the idea of implicit or unconscious knowledge has proved a seductive and controversial topic. Do our minds, as Plato maintained, constantly draw on external "facts" that are akin to recollections? Are there, as some psychologists suggest, discrete systems of thought and calculation that work in concert to increase our mental efficiency but leave us open to error and bias? And if, as Dennis Diderot shrewdly observed, the mind is best envisaged as a book reading itself, who is the book's true author?

The history of what today's psychologists most often term the "cognitive unconscious" began with natural scientists abandoning the soul-based view of human nature. Rejecting the notion that all mental activity was a simple marriage of intellect and will—and taking issue with empiricism's rejection of innate ideas—Enlightenment metaphysicians and natural philosophers found that the mind contained something dark and inherently ungraspable. For Leibniz and Kant, the mystery remained in what lay outside the concentrated beam of conscious thought and perception. But for early evolutionary theorists, the enigma was deeper still: an ancestral thread of unthinking traits that guided human survival and underpinned all our mental life.

As these inquiries into human nature forged new and more complex models of the mind, consciousness was found repeatedly wanting as an explanation of thought and action. Among the league of new experts and specialists who probed this deepening paradox was William Carpenter, originator of the term "unconscious cerebration." One of Victorian

Britian's foremost physiologists, Carpenter proposed that all mental activity was in the first instance spontaneous and automatic, and that instincts, while capable of being enlarged and improved upon, could never be "generated de novo." All thinking was to some degree a form of cerebral reflex action, dictated by forces that we were at best partially aware of.

Carpenter's theory of unconscious cerebration was supported by a small cohort of prominent nineteenth-century thinkers, among them the self-schooled anatomist T. H. Huxley, whose much-quoted essay "On the Hypothesis that Animals are Automata" memorably compared consciousness to the "steam-whistle which accompanies the work of a locomotive engine."[1] Of course, not all of Carpenter's contemporaries agreed that unconscious cerebration was either a useful or necessary concept. The Scottish alienist William Ireland echoed misgivings raised by many British and continental philosophers and physiologists when he argued that Carpenter had wrongly ascribed mental powers to unconscious processes. What this neglected, Ireland insisted, was that the speed of everyday thinking produced a teeming train of associative links that occurred too quickly to be remembered. Our memories were simply not sufficient to keep track of our conscious thoughts and calculations.

For theorists and experimenters who more readily accepted the existence of a fully thinking unconscious mind, questions remained over the practical and moral utility of our quotidian reliance on its secret machinations. In his landmark book *The Philosophy of the Unconscious* (1869), the young German philosopher Eduard von Hartmann, though endorsing the wonderful "clairvoyant" powers of the unconscious mind, could not help but wonder whether knowing judgement and reflection had been signed into a Faustian pact. The same question, when posed by the philosopher and mathematician Alfred North Whitehead, brought a very different conclusion. "It is a profoundly erroneous truism that we should cultivate the habit of thinking of what we are doing," wrote Whitehead. "The precise opposite is the case. Civilization advances by extending the number of operations we can perform without thinking about them."[2]

The idea that willed or conscious thought constitutes only a tiny portion of mental activity, being eclipsed by the vast numbers of operations that we undertake outside of awareness, would go on to become a central tenet of all schools of twentieth century psychology (including psycholinguistics, which Noam Chomsky would eventually lead towards a theory of innate grammar). Finding ways of isolating and measuring mental activity that took place outside the searchlight of consciousness, researchers could now test the extent of incubation and unconscious priming in problem solving; assess the nature of selective and subliminal attention; and discover the heuristics, or mental shortcuts, on which most forms of decision making appeared to rely.

If the upshot of these psychological studies was a model of mind as information-processing system—its filters, buffers, and off-line modes accounting for our capacity to perform multiple tasks, to make snap judgements, or give focused attention to one object or task—a series of experiments undertaken by the American neuroscientist Benjamin Libet, in the early 1980s, went further still in its downgrading of consciousness. Holding that brain activity associated with "readiness potential" preceded the conscious decision to act by around half a second, Libet's widely reported studies arguably confirmed what Huxley had merely intuited a century earlier: free will and volition were the first readers, not true authors, of our thoughts and intentions.

In The Greeks and the Irrational, *Eric R. Dodds (1893–1979), a distinguished scholar of classical literature and philosophy, examined the ancient Greek worldview and challenged the prevailing notion that its mindset was fundamentally rational and empirical.*

ERIC R. DODDS, *THE GREEKS AND THE IRRATIONAL* (1951)

In this matter [of intuitive insight], it seems to me, Plato remained throughout his life faithful to the principles of his master [Socrates]. Knowledge, as distinct from true opinion, remained for him the affair of the intellect, which can justify its beliefs by rational argument. To the intuitions both of the seer and of the poet he consistently refused the title of knowledge, not because he thought them necessarily groundless, but because their grounds could not be produced. Hence Greek custom was right, he thought, in giving the last word in military matters to the commander-in-chief, as a trained expert, and not to the seers who accompanied him on campaign; in general, it was the task of ορθολογική κρίση,

rational judgement, to distinguish between the true seer and the char-
latan. In much the same way, the products of poetic intuition must be
subject to the rational and moral censorship of the trained legislator. All
that was in keeping with Socratic rationalism. Nevertheless, as we have
noticed, Socrates had taken irrational intuition quite seriously, whether
it expressed itself in dreams, in the inner voice of the "daemonion," or
in the utterance of the Pythia. And Plato makes a great show of taking
it seriously too. Of the pseudo-sciences of augury and hepatoscopy he
permits himself to speak with thinly veiled contempt; but "the madness
that comes by divine gift," the madness that inspires the prophet or the
poet, or purges men in the Corybantic rite . . . is treated as if it were a real
intrusion of the supernatural into human life.

[W]hile he thus accepted (with whatever ironical reservations) the
poet, the prophet, and the "Corybantic" as being in some sense channels of divine or daemonic grace, he nevertheless rated their activities
far below those of the rational self, and held that they must be subject
to the control and criticism of reason, since reason was for him no passive plaything of hidden forces, but an active manifestation of deity in
man, a daemon in its own right. I suspect that, had Plato lived to-day, he
would have been profoundly interested in the new depth-psychology,
but appalled by the tendency to reduce the human reason to an instrument for rationalising unconscious impulses.[3]

Considered the father of modern philosophy, Descartes (1596–1650)
introduced a groundbreaking approach to understanding human cognition and
perception. His Meditations on First Philosophy *represents a cornerstone*
in Western thought, the fifth meditation making a case for innate ideas as a
secret engine for insight and learning.

RENÉ DESCARTES, *MEDITATIONS ON FIRST PHILOSOPHY* (1641)

[B]efore investigating whether any such things exist outside me, I should
first consider the ideas of them, in so far as these ideas exist in my thought,
and see which of them are distinct, and which confused.

I can certainly distinctly imagine the quantity that philosophers commonly call 'continuous': that is, the extension of this quantity (or rather,

of the thing to which the quantity is attributed) in length, breadth, and depth. I can count various parts within it. To each of these parts I ascribe various magnitudes, shapes, positions, and local motions, and to the motions I ascribe various durations.

Not only are these things, considered in these general terms, clearly known and grasped by me: I also, if I pay close attention, perceive innumerable particular facts involving shape, number, motion, and suchlike—facts so plainly true, and so much in conformity with my nature, that when I first discover them I do not seem to be learning anything new, but rather to be remembering something I knew before, or to be noticing for the first time something that was in me already, although I had not previously turned the gaze of my mind in its direction.[4]

In opposition to Locke and the empiricists who insisted that the mind was constructed from the building blocks of experience and learning, Leibniz (1646–1716) held that we are "innate to ourselves," and that at any moment our thoughts include an "infinity of perceptions" beyond our conscious grasp.

GOTTFRIED LEIBNIZ, *NEW ESSAYS ON HUMAN UNDERSTANDING* (1765)

Our gifted author [John Locke] seems to claim that there is nothing *implicit* in us, in fact nothing of which we are not always actually aware. But he cannot hold strictly to this; otherwise his position would be too paradoxical, since, again, we are not always aware of our acquired dispositions [*habitude*] or of the contents of our memory, and they do not even come to our aid whenever we need them, though often they come readily to mind when some idle circumstance reminds us of them, as when hearing the opening words of a song is enough to bring back the rest. So on other occasions he limits his thesis to the statement that there is nothing in us of which we have not at least previously been aware. But no one can establish by reason alone how far our past and now perhaps forgotten awarenesses may have extended, especially if we accept the Platonists' doctrine of recollection which, though sheer myth, is entirely consistent with unadorned reason. And furthermore, why must we acquire everything through awareness of outer things and not be able to unearth anything from within ourselves ~ Is our soul in itself so empty that unless it borrows images from outside it is nothing? . . .

I doubt if it will be so easy to make him agree with us and with the Cartesians when he maintains that the mind does not think all the time, and in particular that it has no perceptions during dreamless sleep, arguing that since bodies can be without movement souls can just as well be without thought. . . . [Yet] I maintain that in the natural course of things no substance can lack activity, and indeed that there is never a body without movement. . . . Besides, there are hundreds of indications leading us to conclude that at every moment there is in us an infinity of perceptions, unaccompanied by awareness or reflection; that is, of alterations in the soul itself, of which we are unaware because these impressions are either too minute and too numerous, or else too unvarying, so that they are not sufficiently distinctive on their own.[5]

Greatly influenced by Leibniz, Immanuel Kant's (1724–1804) notion of a perceptual unconscious—of seeing more than we apprehend—became a touchstone for transcendental idealism.

IMMANUEL KANT, *ANTHROPOLOGY FROM A PRAGMATIC POINT OF VIEW* (1798)

A contradiction appears to lie in the claim to have representations and still not be conscious of them? Locke already raised this objection, and this is why he also rejected the existence of representations of this nature; however, we can still be *indirectly* conscious of having a representation, even if we are not directly conscious of it. Such representations are then called *obscure*; the others are *clear*, and when their clarity also extends to the partial representations that make up a whole together with their connection, they are then called *distinct representations*, whether of thought or intuition.

When I am conscious of seeing a human being far from me in a meadow, even though I am not conscious of seeing his eyes, nose, mouth, etc., I properly *conclude* only that this thing is a human being. For if I wanted to maintain that I do not at all have the representation of him in my intuition because I am not conscious of perceiving these parts of his head (and so also the remaining parts of this human being), then I would also not be able to say that I see a human being, since the representation of the whole (of the head or of the human being) is composed of these partial ideas.

The field of sensuous intuitions and sensations of which we are not conscious, even though we can undoubtedly conclude that we have them; that is, *obscure* representations in the human being (and thus also in animals), is immense. Clear representations, on the other hand, contain only infinitely few points of this field which lie open to consciousness; so that as it were only a few places on the vast *map* of our mind are *illuminated.*[6]

Franz Gall (1758–1828), best known as the pioneer of the popular science of phrenology, believed that both moral and intellectual dispositions were innate in both animals and humans, their manifestation depending on their "organization" within the brain.

FRANZ GALL, *ON THE FUNCTION OF THE BRAIN AND OF EACH OF ITS PARTS* (1825)

Why does not the dog build a house to protect him from the inclemencies of the weather? Why do the partridge and raven perish of cold, rather than migrate like the swallow? Why is it that each animal satisfies its wants in a manner peculiar to itself? That each man has different wants, though outward circumstances are very nearly the same in all? In treating of the fundamental powers of man and brute, I have satisfactorily answered these questions. The true source of the arts and sciences is our innate instincts, propensities, and faculties—our inward wants.

Who invented the spider's web, the beaver's cabin, the hang-bird's nest, the bee's cells, the nightingale's song? Who suggested the idea of a republic to the ants, of tricks to the monkeys, of sentinels to the chamois, of migration to the storks, of hunting to the wolves, of provisions to the hamsters, of marriage to nearly all the birds, and a large part of the mammifera? All these things are universally attributed to instinct, to an inward impulse, never to external circumstances. The cause of these inventions, therefore, lies in the organs, or, in other words, animals have received from nature, by means of organs, certain definite powers, pro-pensities, talents, and faculties, which produce their habits, that have so often the appearance of spontaneous and deliberate actions. It is precisely the same with man. All that he does, or knows, all that he can do, or can learn, he owes to the author of his organization. God is its source; the cerebral organs, his intermediate instrument.[7]

In the years leading up to his publication of The Origin of Species, *Charles Darwin (1809–1882) recorded his evolving ideas on natural selection and the origins of human cognition in a series of private notebooks, trailing the idea that complex traits, including mental faculties, could be naturally inherited.*

CHARLES DARWIN, "M NOTEBOOKS" (1838)

Plato says in Phaedo that our "necessary ideas" arise from the preexis-tence of the soul, are not derivable from experience—read monkeys for preexistence.[8]

Anticipating certain aspects of Darwinian theory, the German philosopher Arthur Schopenhauer (1788–1860) is widely regarded as the first modern philosopher of the unconscious.

ARTHUR SCHOPENHAUER, "SOME THOUGHTS CONCERNING THE INTELLECT" (1851)

Most often the conclusion comes without the premises having been clearly thought out. This can already be inferred from the fact that occa-sionally an event whose consequences we do not foresee at all, and whose possible influence on our own affairs we cannot clearly measure, none-theless exerts an undeniable influence on our whole mood by changing it to cheerful or to sad. This can only be the result of an unconscious rumi-nation, which is more evident in the following. I have familiarized myself with the factual data of a theoretical or practical matter; often the result, for instance the disposition of the matter or what is to be done about it, comes to me entirely on its own and stands plainly in front of me after a few days without my having thought about it again. Meanwhile the operation according to which this has come about remains as hidden to me as the workings of a calculator; for it was nothing but an unconscious rumination. In the same manner, when I have recently written something on a theme but then let the matter drop, occasionally something more on it will occur to me even though I was not thinking about it at all in the meantime. Similarly, I can search in my memory for days for a name that I have forgotten, but then when I am not thinking about it in the least it suddenly occurs to me as if whispered in my ear. Indeed, our best, most ingenious and profound thoughts enter suddenly into consciousness like an inspiration and often even in the form of a weighty sentence. But obviously they are the results of long, unconscious meditation and countless *aperçus* extending far back and individually forgotten.[9]

Straddling aesthetics and psychology, The Gay Science *was the first work of Victorian literary criticism to draw on the new theories of unconscious cerebration.*

ENEAS SWEETLAND DALLAS, *THE GAY SCIENCE* (1866)

[N]obody tells us what imagination really is, and how it happens that being, as some say, nothing at all, plays an all-powerful part in human life. Driven to our own resources, we must see if we cannot give a clearer account of this wonder-working energy, and above all, cannot reconcile the philosophical analysis which reduces imagination to a shadow with the popular belief which gives it the empire of the mind. I propose this theory, that the imagination or fantasy is not a special faculty but that it is a special function. It is a name given to the automatic action of the mind or any of its faculties—to what may not unfitly be called the Hidden Soul. . . .

Now for the most part this automatic action takes place unawares; and when we come to analyse the movements of thought we thought we find to be quite sure of our steps we are obliged very much to identify what is involuntary with what is unconscious. We are seldom quite sure that our wills have had nought to do in producing certain actions, unless these actions have come about without our knowledge. Therefore although involuntary does not in strictness coincide with unconscious action, yet for practical purposes, and, above all, for the sake of clearness, it may be well to put out of sight altogether such involuntary action as may consist with full consciousness, and to treat of the automatic exercise of the mind as either quite unconscious or but half conscious. And if on this understanding we may substitute the one phrase for the other as very nearly coinciding, then the task before me is to show that imagination is but a name for the unknown, unconscious action of the mind—the whole mind or any of its faculties—for the Hidden Soul. . . .

This unconscious part of the mind is so dark, and yet so full of activity; so like the conscious intelligence and yet so divided from it by the veil of mystery, that it is not much of a hyperbole to speak of the human soul as double; or at least as leading a double life.[10]

Eduard von Hartmann's (1842–1906) Philosophy of the Unconscious *was the first best-selling work on the powers of the unconscious mind. While celebrating the personal and cosmic reach of the unconscious, von Hartmann served a warning about surrendering to the gifts of intuition.*

EDUARD VON HARTMANN, *PHILOSOPHY OF THE UNCONSCIOUS* (1868)

[E]verything which any consciousness has power to accomplish can be executed equally well by the Unconscious, and that too always far more strikingly, and therewith more quickly and more conveniently for the individual, since the conscious performance must be striven for, whereas the Unconscious comes of itself and without effort. This convenience of abandoning oneself to the Unconscious, its feelings and inspirations, is tolerably familiar, and hence the conscious use of reason is so decried in all and every one by the indolent. That the Unconscious can really outdo all the performances of conscious reason, is what we should not only à priori expect of the clairvoyance of the Unconscious, but we see it also realised in those fortunate natures which possess everything that others must acquire with toil, who never have a struggle of conscience, because they always spontaneously act correctly and morally in accordance with feeling, can never comport themselves otherwise than with tact, learn everything easily, complete everything which they begin with a happy

knack, and live in eternal harmony with themselves, without ever reflecting much what they do, or even experiencing difficulty and toil. . . .

But now what disadvantage lies in this self-surrender to the Unconscious? This, that one never knows where one is or what one has; that
one gropes in the dark, while one has got the lantern of consciousness
in one's pocket; that it is left to accident, whether the inspiration of the
Unconscious will come when one wants it; that one has no criterion but
success, what is an inspiration of the Unconscious, and what a wrong-
headed flash of whimsical fancy, on what feeling one may rely, and on
what not; finally, that one does not practise conscious judgment and
reflection, which can never be entirely dispensed with, and that then in
any case which occurs one must put up with wretched analogies instead
of rational inferences and all-sided survey. Only the Conscious one
knows as one's own, the Unconscious confronts us as something incomprehensible, foreign, on whose favour we are dependent; the Conscious
is possessed as ever-ready servant, whose obedience may be always compelled; the Unconscious protects us like a fairy, and has always something
uncomfortably demonic about it.[11]

In The Origin of Species *(1859), Darwin conceded that "many instincts
are so wonderful that their development will probably appear to the reader
a difficulty sufficient to overthrow my whole theory." Enlarging his theory
of natural selection, Darwin went on to catalogue a host of psychological
instincts that operated alongside those of survival and reproduction, including
the most human of sentiments—sympathy.*

CHARLES DARWIN, *THE EXPRESSION OF EMOTIONS IN MAN AND ANIMALS* (1872)

As most of the movements of expression must have been gradually acquired,
afterwards becoming instinctive, there seems to be some degree of a priori
probability that their recognition would likewise have become instinctive.
There is, at least, no greater difficulty in believing this than in admitting
that, when a female quadruped first bears young, she knows the cry of distress of her offspring, or than in admitting that many animals instinctively
recognize and fear their enemies; and of both these statements there can

be no reasonable doubt. It is however extremely difficult to prove that our children instinctively recognize any expression. I attended to this point in my first-born infant, who could not have learnt anything by associating with other children, and I was convinced that he understood a smile and received pleasure from seeing one, answering it by another, at much too early an age to have learnt anything by experience. When this child was about four months old, I made in his presence many odd noises and strange grimaces, and tried to look savage; but the noises, if not too loud, as well as the grimaces, were all taken as good jokes; and I attributed thus at the time to their being preceded or accompanied by smiles. When five months old, he seemed to understand a compassionate expression and tone of voice. When a few days over six months old, his nurse pretended to cry, and I saw that his face instantly assumed a melancholy expression, with the corners of the mouth strongly depressed; now this child could rarely have seen any other child crying, and never a grown-up person crying, and I should doubt whether at so early an age he could have reasoned on the subject. Therefore it seems to me that an innate feeling must have told him that the pretended crying of his nurse expressed grief; and this through the instinct of sympathy excited grief in him.[12]

Physiological research into reflex actions and unconscious cerebration soon dovetailed with the new evolutionary theories of instinct. Alongside Thomas Laycock, the physiologist William Benjamin Carpenter (1813–1885) was at the forefront of investigations into what is today often termed the adaptive unconscious.

WILLIAM BENJAMIN CARPENTER, *PRINCIPLES OF MENTAL PHYSIOLOGY* (1874)

Having thus found reason to conclude that a large part of our Intellectual activity—whether it consist in Reasoning processes, or in the exercise of the Imagination—is essentially *automatic*, and may be described in Physiological language as the *reflex action of the Cerebrum*, we have next to consider whether this action may not take place *unconsciously*. To affirm that the Cerebrum may act upon impressions transmitted to it, and may elaborate Intellectual results, such as we might have attained by the intentional direction of our minds to the subject, *without any conscious-ness* on our own parts, is held by many Metaphysicians, more especially in Britain, to be an altogether untenable and even a most objectionable doctrine. But this affirmation is only the Physiological expression of a doctrine which has been current among the Metaphysicians of Germany, from the time of Leibniz to the present date, and which was systemati-cally expounded by Sir William Hamilton,—that the Mind may undergo modifications, sometimes of very considerable importance, without being itself conscious of the process, until its *results* present themselves to the consciousness, in the new ideas, or new combinations of ideas, which the process has evolved. This "Unconscious Cerebration," or "Latent Mental modification," is the precise parallel, in the higher sphere of Cerebral or Mental activity, to the movements of our limbs, and to the direction of those movements through our visual sense, which we *put in train* voli-tionally when we set out on some habitually repeated walk, but which then proceed not only *automatically*, but *unconsciously*, so long as our attention continues to be uninterruptedly diverted from them.[13]

Often remembered as "the father of eugenics," the British polymath Francis Galton's (1822–1911) limitless curiosity led him to explore various aspects of human psychology, including the automaticity of thinking.

FRANCIS GALTON, *INQUIRIES INTO HUMAN FACULTY AND ITS DEVELOPMENT* (1883)

When I am engaged in trying to think anything out, the process of doing so appears to me to be this: The ideas that lie at any moment within my full consciousness seem to attract of their own accord the most appro-priate out of a number of other ideas that are lying close at hand, but imperfectly within the range of my consciousness. There seems to be a presence-chamber in my mind where full consciousness holds court, and where two or three ideas are at the same time in audience, and an ante-chamber full of more or less allied ideas, which is situated just beyond the full ken of consciousness. Out of this antechamber the ideas most nearly allied to those in the presence-chamber appear to be summoned in a mechanically logical way, and to have their turn of audience.

The successful progress of thought appears to depend—first, on a large attendance in the antechamber; secondly, on the presence there of no ideas except such as are strictly germane to the topic under consideration; thirdly, on the justness of the logical mechanism that issues the summons. The thronging of the antechamber is, I am convinced, altogether beyond my control; if the ideas do not appear, I cannot create them, nor compel them to come. The exclusion of alien ideas is accompanied by a sense of mental effort and volition whenever the topic under consideration is unattractive, otherwise it proceeds automatically, for if an intruding idea finds nothing to cling to, it is unable to hold its place in the antechamber, and slides back again.[14]

Recognizing the depth and range of unthinking instincts and impulses, William James (1842–1910) stressed that this did not render us "fatal automatons."

WILLIAM JAMES, "WHAT IS AN INSTINCT?" (1887)

Man has a far greater variety of *impulses* than any lower animal; and any one of these impulses, taken in itself, is as "blind" as the lowest instinct can be; but, owing to man's memory, power of reflection, and power of inference, they come each one to be felt by him, after he has once yielded to them and experienced their results, in connection with a *foresight* of those results. In this condition an impulse acted out may be said to be acted out, in part at least, *for the sake* of its results. It is obvious that every instinctive act, in an animal with memory, must cease to be "blind" after being once repeated, and must be accompanied with foresight of its "end" just so far as that end may have fallen under the animal's cognizance. An insect that lays her eggs in a place where she never sees them hatch must always do so "blindly;" but a hen who has already hatched a brood can hardly be assumed to sit with perfect "blindness" on her second nest. Some expectation of consequences must in every case like this be aroused; and this expectation, according as it is that of something desired or of something disliked, must necessarily either re-enforce or inhibit the mere impulse. . . .

It is plain then that, no matter how well endowed an animal may originally be in the way of instincts, his resultant *actions* will be much modified if the instincts *combine* with experience, if in addition to impulses he have memories, associations, inferences, and expectations, on any considerable scale. An object O, on which he has an instinctive impulse to react in the manner A, would *directly* provoke him to that reaction. But O has meantime become for him a *sign* of the nearness of P, on which he has an equally strong impulse to react in the manner B, quite unlike A. So that when he meets O the immediate impulse A and the remote impulse B struggle in his breast for the mastery. The fatality and uniformity said to be characteristic of instinctive actions are so little manifest, that one might be tempted to deny to him altogether the possession of any instinct about the object O. Yet how false this judgment would be! The instinct about O is there; only by the complication of the mental machinery it has come into conflict with another instinct about P. . . .

Thus, then, without troubling ourselves about the words instinct and reason, we may confidently say that however uncertain man's reactions upon his environment may sometimes seem in comparison with those of lower creatures, the uncertainty is probably not due to their possession of any principles of action which he lacks, but to his possessing all the impulses that they have, and a great many more besides. In other words, there is no material antagonism between instinct and reason. Reason, per se, can inhibit no impulses; the only thing that can neutralize an impulse is an impulse the other way. Reason may, however, make an *inference which will set loose* the impulse the other way; and thus, though the animal richest in reason might be also the animal richest in instinctive impulses too, he would never seem the fatal automaton which a *merely* instinctive animal would be.[15]

Yet most intellectual reasoning was, according to the American psychologist Joseph Jastrow (1863–1944), neither controlled nor conscious.

JOSEPH JASTROW, *THE SUBCONSCIOUS* (1906)

There exists in all intellectual endeavor a period of incubation, a process in great part subconscious, a slow, concealed maturing through absorption of suitable pabulum. Schopenhauer calls it "unconscious rumination," a chewing over and over again of the cud of thought preparatory to its assimilation with our mental tissue; another speaks of it as the red glow that precedes the white heat. The thesis implied by such terms has two aspects: first, that the process of assimilation may take place with suppressed consciousness; second, that the larger part of the influences that in the end determine our mental growth may be effective without direct exposure to the searching light of conscious life. Both principles enforce the view that we develop by living in an atmosphere congenial to the occupation that we seek to make our own; by steeping ourselves in the details of the business that is to be our specialty, until the judgement is trained, the assimilation sensitized, the perspective of importance for the special purpose well established, the keenness for useful improvisation brought to an edge. When asked how he came to discover the law of gravitation, Newton is reported to have answered, "By always thinking about it."

While the second aspect of this thesis is hardly susceptible of any more definite illustration than is afforded by the general cultural fruitage of our combined nature and nurture, the first aspect presents a precise problem, which the psychologist approaches with such special equipment as his ingenuity affords. His method is to catch the moment of perception at the lapsing edge of consciousness and forcibly to reinstate it; for there is an area in which, under favorable circumstances, the passage in and out of the range of the inner search-light may be rendered visible. There is, for instance, the common experience that something which we were just ready to speak has, by the rivalry of other intruded interests, been temporarily driven back from consciousness, and leaves us adrift, the conscious vacantly asking the subconscious self, "What was I going to say?" It is by a sort of fumbling about among the fading trails of ideas for some clue by which to recover the lost thread of discourse, that we attempt to arrest the fast receding lines of thought. A variation of this experience occurs in writing, whenever a larger group of suggested ideas than can immediately find expression appeals for notice; the writer has the troubled feeling that, while recording one, the others will again slip from his mental grasp. In all original composition there occur constant relaxations in the tension of thought—at times the budding of a brief abstraction—in which the associations that had just entered the focus of awareness flit back into the shadow and must again be sought for when the light of attention in turn brightens. The very attitude of the effort to recover such evasive associations—the closing of the eyes to exclude the outer glare and relieve by contrast the dimness of the light within, the intent peering in the dark to catch the first glimmer of the lost trail—is suggestive of the procedure which the mind may be said figuratively to employ. In such wise may we occasionally detect the exit of ideas hovering near the margins of consciousness, when our interest makes us eager for their recovery. Frequently do we fail in this endeavor, the failure inducing a submerged troubled feeling while the mental explorer goes forth and "comes back like the dove into the ark, having found no rest;" and we either make the attempt anew under more promising auspices, or are agreeably surprised by the spontaneous intrusion of the lost idea into our otherwise occupied attention.[16]

Regarding the intellect as being dulled by "a natural inability to comprehend life," Henri Bergson's (1859–1941) highly popular book Creative Evolution *found that our instincts were a mute form of vital intelligence.*

HENRI BERGSON, *CREATIVE EVOLUTION* (1907)

While intelligence treats everything mechanically, instinct proceeds, so to speak, organically. If the consciousness that slumbers in it should awake, if it were wound up into knowledge instead of being wound off into action, if we could ask and it could reply, it would give up to us the most intimate secrets of life. For it only carries out further the work by which life organizes matter so that we cannot say, as has often been shown, where organization ends and where instinct begins. When the little chick is breaking its shell with a peck of its beak, it is acting by instinct, and yet it does but carry on the movement which has borne it through embryonic life. Inversely, in the course of embryonic life itself (especially when the embryo lives freely in the form of a larva), many of the acts accomplished must be referred to instinct. The most essential of the primary instincts are really, therefore, vital processes. The potential consciousness that accompanies them is generally actualized only at the outset of the act, and leaves the rest of the process to go on by itself.

It would only have to expand more widely, and then dive into its own depth completely, to be one with the generative force of life.[17]

During World War One, the psychologist Wolfgang Köhler (1887–1967) directed the German Primate Research Station on the island of Tenerife. During his landmark studies with captive chimpanzees, Köhler found that his stand-out performer, Sultan, often solved food-gathering problems through sudden insight rather than trial-and-error learning. Thanks to Kohler, the notion of Einsicht (insight) would become one of the central pillars of Gestalt psychology.

WOLFGANG KÖHLER, *THE MENTALITY OF APES* (1917)

Beyond the bars lies the objective [a bunch of bananas]. . . .

Sultan first of all squats indifferently on the box, which has been left standing a little back from the railings; then he gets up, picks up the two sticks, sits down again on the box and plays carelessly with them. While doing this, it happens that he finds himself holding one rod in either hand in such a way that they lie in a straight line; he pushes the thinner one a little way into the opening of the thicker, jumps up and is already on the run towards the railings, to which he has up to now half turned his back, and begins to draw a banana towards him with the double stick. . . .

Sultan is squatting at the bars, holding out one stick, and, at its end, a second bigger one, which is on the point of falling off. It does fall. Sultan pulls it to him and forthwith, with the greatest assurance, pushes the thinner one in again, so that it is firmly wedged, and fetches a fruit with the lengthened implement. But the bigger tube selected is a little too big, and so it slips from the end of the thinner one several times; each time Sultan rejoins the tubes immediately by holding the bigger one towards himself in the left and the thinner one in his right hand and a little backwards, and then sliding one into the other. The proceeding seems to please him immensely; he is very lively, pulls all the fruit, one after the other, towards the railings, without taking time to eat it, and when I disconnect the double-stick he puts it together again at once, and draws any distant objects whatever to the bars.[18]

*The growth of popular psychology in the early decades of the twentieth century
can be charted through a heaving bookshelf of works on the art and science
of thinking and mental fitness. In* The Art of Thinking, *Graham Wallas
(1858–1932), psychologist and co-founder of the London School of Economics,
proposed that sudden illuminations were rarely sudden or self-evident.*

GRAHAM WALLAS, *THE ART OF THOUGHT* (1926)

An economist reading a Blue Book, a physiologist watching an experiment, or a business man going through his morning's letters, may at the same time be 'incubating' on a problem which he proposed to himself a few days ago, be accumulating knowledge in 'preparation' for a second problem, and be 'verifying' his conclusions on a third problem. Even in exploring the same problem, the mind may be unconsciously incubating on one aspect of it, while it is consciously employed in preparing for or verifying another aspect. And it must always be remembered that much very important thinking, done for instance by a poet exploring his own memories, or by a man trying to see clearly his emotional relation to his country or party, resembles musical composition in that the stages leading to success are not very easily fitted into a 'problem and solution' scheme. Yet, even when success in thought means the creation of something felt to be beautiful and true rather than the solution of a prescribed problem, the four stages of Preparation, Incubation, Illumination, and the Verification of the final result can be generally distinguished from each other.[19]

*Even the hapless Bertie Wooster, ever reliant on his valet Jeeves to resolve his
social dilemmas, was now au fait with the problem-solving powers of "the
subconscious mind."*

P. G. WODEHOUSE, *RIGHT HO, JEEVES* (1934)

I don't know if it has happened to you at all, but a thing I've noticed with myself is that, when I'm confronted by a problem which seems for the moment to stump and baffle, a good sleep will often bring the solution in the morning.

It was so on the present occasion.

The nibs who study these matters claim, I believe, that this has got something to do with the subconscious mind, and very possibly they may be right. I wouldn't have said off-hand that I had a subconscious mind, but I suppose I must without knowing it, and no doubt it was there, sweating away diligently at the old stand, all the while the corporeal Wooster was getting his eight hours.

For directly I opened my eyes on the morrow, I saw daylight. Well, I don't mean that exactly, because naturally I did. What I mean is that I found I had the thing all mapped out. The good old subconscious m. had delivered the goods, and I perceived exactly what steps must be taken in order to put Augustus Fink-Nottle among the practising Romeos.[20]

The British psychologist William McDougall (1871–1938), a long-time nemesis of behaviorism, championed what he called hormic or purposive psychology, identifying up to eighteen instincts that were drivers of human conduct.

WILLIAM MCDOUGALL, *AN INTRODUCTION TO SOCIAL PSYCHOLOGY* (1936)

Man also is a member of an animal species. And this species also has its natural goals, or its inborn tendencies to seek goals of certain types. This fact is not only indicated very clearly by any comparison of human with animal behaviour, but it is so obvious a fact that no psychologist of the least intelligence fails to recognise it, however inadequately, not even if he obstinately reduces their number to a minimum of three and dubs them the "prepotent reflexes" of sex, fear, and rage. Others write of "primary desires," or of "dominant urges," or of "unconditioned reflexes," or of appetites, or of cravings, or of congenital drives, or of motor sets, or of inherited tendencies or propensities; lastly, some, bolder than the rest, write of "so-called instincts."—For instincts are out of fashion just now with American psychologists; and to write of instincts without some such qualification as "so-called" betrays a reckless indifference to fashion amounting almost to indecency. Yet the word "instinct" is too good to be lost to our science. Better than any other word it points to the facts and the problems with which I am here concerned.

The hormic psychology imperatively requires recognition not only of instinctive action but of instincts. Primarily and traditionally the words "instinct" and "instinctive" point to those types of animal action which are complex activities of the whole organism; which lead the creature to the attainment of one or other of the goals natural to the species; which are in their general nature manifested by all members of the species under appropriate circumstances; which exhibit nice adaptation to circumstances; and which, though often suggesting intelligent appreciation of the end to be gained and the means to be adopted, yet owe little or nothing to the individual's prior experience.[21]

As part of his research on the roots of creativity in the field of mathematics, Jacques Hadamard (1865–1963) wrote to a host of scientific luminaries asking them to describe their working methods. Foremost among Hadamard's correspondents was Albert Einstein.

ALBERT EINSTEIN, LETTER TO JACQUES HADAMARD (1945)

The words or the language, as they are written or spoken, do not seem to play any role in my mechanism of thought. The psychical entities which seem to serve as elements in thought are certain signs and more or less clear images which can be "voluntarily" reproduced and combined.

There is, of course, a certain connection between those elements and relevant logical concepts. It is also clear that the desire to arrive finally at logically connected concepts is the emotional basis of rather vague play with the above mentioned elements. But taken from a psychological viewpoint, this combinatory play seems to be the essential feature in productive thought before there is any connection with logical construction in words or other kinds of signs which can be communicated to others. . . .

It seems to me that what you call full consciousness is a limit case which can never be fully accomplished. This seems to me connected with the fact called the narrowness of consciousness (*Enge des Bewusstseins*).[22]

The Austrian ethologist Konrad Lorenz (1903–1989) seconded the views of
Lorenz and the Gestaltists, stressing the importance of intuition's unconscious
computations for science as well as survival.

KONRAD LORENZ, *THE ROLE OF GESTALT PERCEPTION IN ANIMAL AND HUMAN BEHAVIOUR* (1951)

Intuition is generally regarded as the prerogative of artists and poets. I would assert that it plays an indispensable role in all human recognition, even in the most disciplined forms of inductive research. Though in the latter the important part taken by intuition is very frequently overlooked, no important scientific fact has ever been "proved" that had not previously been simply and immediately seen by intuitive Gestalt perception. Intuition it was when Kepler first perceived, in the complicated epicycles of the planets' apparent movements, the simple regularity of their real orbits, or when Darwin first saw, in the intricate tangle of living and extinct forms of life, the convincingly clear Gestalt of the genealogical tree. Without intuition, the world would present to us nothing but an impenetrable and chaotic tangle of unconnected facts. It would be quite impossible to us to find the laws and regularities prevailing in this apparent chaos, if the mathematical and statistical operations of our conscious mind were all that we had at our disposal. It is here that the unconsciously working computor of our Gestalt perception is distinctly superior to all consciously performed computations.

This superiority is due to the fact that intuition, like other highly differentiated types of Gestalt perception, is able to draw into simultaneous consideration a far greater number of premises than any of our conscious conclusions. It is the practically unlimited capacity for taking in relevant details and leaving out the irrelevant ones which makes the computor of this highest form or Gestalt perception so immensely sensitive an organ. The most important advantage of intuition is that it is "seeing" in the deepest sense of the word. Like other kinds of Gestalt perception and unlike inductive research, it does not only find what is expected, but totally unexpected as well.[23]

Drawing heavily on Gestalt psychology, Arthur Koestler (1905–1983) began to investigate the conditions that led to the solving of creative problems in his 1964 book The Act of Creation, *in which he coined the term "bisociation."*

ARTHUR KOESTLER, "CREATIVITY AND THE UNCONSCIOUS" (1980)

When the challenge cannot be met, the problem is *blocked*—though the subject may realise this only after a series of hopeless tries, or never at all. A blocked situation increases the stress of the frustrated drive. When all promising attempts at solving a problem by traditional means have been exhausted, thinking tends to run in circles like cats in the puzzle-box, until the whole person becomes saturated with the problem. At this stage—the 'period of incubation'—the single-mindedness of the creative obsession produces a state of receptivity, a readiness to pounce on favour-able chance constellations and to profit from any casual hint. As Lloyd Morgan said: 'Saturate yourself through and through with your subject, and wait.' Thus in discoveries of the type in which both rational thinking and the trigger-action of chance play a noticeable part, the main contri-bution of the unconscious is to keep the problem on the agenda while conscious attention is occupied elsewhere—like Newton watching an apple fall. . . .

But in other types of discovery unconscious mentation seems to inter-vene in more specific, active ways, and enable the mind to perform sur-prisingly original, quasi-acrobatic feats, which leads to revolutionary breakthroughs in science or art, open new vistas, and create a radically changed outlook. . . . [This] always involves un-learning and re-learning, undoing and re-doing. It involves the breaking up of petrified mental structures, discarding matrices which have outlived their usefulness, and reassembling others in a new synthesis—in other words, it is a complex operation of *dissociation* and *bisociation*, involving several labels of mental hierarchy.[24]

The linguist Noam Chomsky (1928–) proposed that we are all born with a readymade capacity for language—a "Language Acquisition Device" that

allows us to intuitively and unconsciously grasp the rules and structures of
language.

NOAM CHOMSKY, *REFLECTIONS ON LANGUAGE* (1975)

No one would take seriously a proposal that the human organism learns through experience to have arms rather than wings, or that the basic structure of particular organs results from accidental experience. Rather, it is taken for granted that the physical structure of the organism is genetically determined, though of course variation along such dimensions as size, rate of development, and so forth will depend in part on external factors. From embryo to mature organism, a certain pattern of development is predetermined, with certain stages, such as the onset of puberty or the termination of growth, delayed by many years. Variety within these fixed patterns may be of great importance for human life, but the basic questions of scientific interest have to do with the fundamental, genetically determined scheme of growth and development that is a characteristic of the species and that gives rise to structures of marvelous intricacy. . . .

The development of personality, behavior patterns, and cognitive structures in higher organisms has often been approached in a very different way. It is generally assumed that in these domains, social environment is the dominant factor. The structures of mind that develop over time are taken to be arbitrary and accidental; there is no "human nature" apart from what develops as a specific historical product. . . . But human cognitive systems, when seriously investigated, prove to be no less marvelous and intricate than the physical structures that develop in the life of the organism. Why, then, should we not study the acquisition of a cognitive structure such as language more or less as we study some complex bodily organ?

At first glance, the proposal may seem absurd, if only because of the great variety of human languages. But a closer consideration dispels these doubts. Even knowing very little of substance about linguistic universals, we can be quite sure that the possible variety of languages is sharply limited. Gross observations suffice to establish some qualitative conclusions. Thus, it is clear that the language each person acquires is a rich and complex construction hopelessly underdetermined by the fragmentary evidence available. . . . [That] individuals in a speech community have developed essentially the same language . . . can be explained only on the assumption that these individuals employ highly restrictive principles that guide the construction of grammar.[25]

One of the most significant of the "deep structures" or innate modules to be proposed in the aftermath of Chomsky's universal grammar was "a theory of mind," the ability to read others' goals and intentions through their outward behavior.

DAVID PREMACK AND GUY WOODRUFF, "DOES THE CHIMPANZEE HAVE A THEORY OF MIND?" (1978)

In assuming that other individuals want, think, believe, and the like, one infers states that are not directly observable and one uses these states anticipatorily, to predict the behavior of others as well as one's own. These inferences, which amount to a theory of mind, are, to our knowledge, universal in human adults. Although it is reasonable to assume that their occurrence depends on some form of experience, that form is not immediately apparent. Evidently it is not that of an explicit pedagogy. Inferences about another individual are not taught, as are reading or arithmetic; their acquisition is more reminiscent of that of walking or speech. Indeed, the only direct impact of pedagogy on these inferences would appear to be suppressive, for it is only the specially trained adult who can give an account of human behavior that does not impute states of mind to the participants. All this is to say that theory building of this kind is natural in man.

Are we to believe, however, that we are the only species in which it is natural? . . . if mental theories are indeed natural, this fact must have untoward consequences for behaviorism. After being shown that not only man but also apes had theories of mind, suppose a behaviorist were to reply, "Yes . . . and they are both wrong." Would this save behaviorism? We think not, for to admit that animals are mentalists compels the admission that behaviorist accounts of animals are at best profoundly incomplete. Moreover—and we add this with more than facetious intent—it would waste the behaviorist's time to recommend parsimony to the ape.

The ape could only be a mentalist. Unless we are badly mistaken, he is not intelligent enough to be a behaviorist.[26]

The phenomenon of blindsight, first studied by the British psychologist Larry Weiskrantz (1926–2018), revealed that visual perception also had multiple pathways, some of which operated outside of conscious awareness.

NICHOLAS HUMPHREY, *THE INNER EYE* (1986)

Blindsight is seeing without knowing that you can see; unconscious vision; seeing things which to your conscious mind are quite invisible. In 1974 a patient, known by the initials D.B., was examined by Professor Weiskrantz and his colleagues at the National Hospital in London. D.B. had recently undergone surgery to remove a growth at the back of his brain—an operation which meant the excision of the entire primary visual cortex on the right-hand side. The effect of the lesion was, as predicted from earlier clinical studies, that D.B. became blind in the left side of his field of vision. When, for example, he looked straight ahead he could not (with either eye) see anything to the left of his nose. Or so it seemed, both to D.B. himself and to the doctors who first tested his vision by asking him to tell them whether he could see a light going on or off in different parts of the field.

Weiskrantz, however, decided not to accept D.B.'s self-professed blindness at face value. There was no question that D.B. was genuinely unaware of seeing anything in the blind half of his field; but was it possible that his brain none the less was still receiving and processing the visual information? What would happen if he could be persuaded to discount his own conscious opinion?

Weiskrantz asked D.B. to forget for a moment that he was blind, and to 'guess' at what he might be seeing if he could see. To D. B's own amazement, it turned out that he could do it. He could locate an object accurately in his blind field, and he could even guess certain aspects of its shape. Yet all the while he denied any conscious awareness.

Other cases of human 'blindsight' have now been described. . . . Consciousness, it seems, is not necessary to perception. Not only animals, but human beings themselves, can function without it, which opens up a huge range of possibilities. For if we can perceive without being conscious of any accompanying sensations, why should we not think without being conscious of our thoughts, act without being conscious of our own intentions, even be ourselves without, in any real sense, feeling we exist?[27]

Are we strangers to our own intentions? Are our free and conscious decisions the upshot of inaccessible deliberations? In the early 1980s, Benjamin Libet's (1916–2007) classic experiments at the University of California appeared to suggest that, as Schopenhauer and others had suggested, free will is merely reactive.

BENJAMIN LIBET, *MIND TIME* (2004)

Our experimental objective was to study freely voluntary acts, performed with no external restrictions as to when to act. In most of our series, each of forty trials, there were no reports of preplanning by the subjects. These voluntary acts were completely free and performed spontaneously, without any preplanning of when to act. The nature of the act, sudden flexion of the wrist, was of course prescribed by us for the subject. That allowed us to place recording electrodes on the actual muscle to be activated; the recorded electromyogram gave us the time of the act and also served as a trigger to the computer to record the scalp potential that had appeared during the 2 to 3 sec prior to the muscle activation. But the time of the act was completely free for the subjects' own will. Our experimental question was: Does the conscious will to act precede or follow the brain's action? . . .

The clear answer was: The brain initiates the voluntary process first. The subject later becomes consciously aware of the urge or wish (W) to act, some 350 to 400 msec after the onset of the recorded RP produced by the brain. This was true for every series of forty trials with every one of the nine subjects. . . .

These results lead to a different way of regarding the role of conscious will and of free will, in a volitional process leading to an act. Extrapolating our result to other voluntary acts, conscious free will does not initiate our freely voluntary acts. Instead, it can control the outcome or actual performance of the act. It could permit the action to proceed, or it can veto it, so that no action occurs.[28]

Nancy Andreasen (1938–) began to conduct laboratory studies of creativity in the early 1970s, going on to find that creative thinking had its own special neural signature.

NANCY C. ANDREASEN, *THE CREATING BRAIN* (2005)

One may conclude not only that extraordinary creativity is at least sometimes based on a qualitatively different neural process than ordinary creativity, but that it at least sometimes arises from that "over the precipice" component of human thought that we call the unconscious. What is that process, and how does it arise? We have seen . . . that free-floating and uncensored thought (primary-process thought, primitive thought, original thought) occurs when multiple regions of our highly developed human association cortex interact with one another. When this occurs, the brain is working as a self-organizing system, but in a different way. Think of Poincaré's impression that "ideas rose in crowds; I felt them collide until pairs interlocked, so to speak, making a stable combination." These introspective accounts are describing a process during which thought is not only nonsequential or nonlinear, but during which nonrational unconscious processes play a role. It is as if the multiple association cortices are communicating back and forth, not in order to integrate associations with sensory or motor input as is often the case, but simply in response to one another. The associations are occurring freely. They are running unchecked, not subject to any of the reality principles that normally govern them. Initially these associations may seem meaningless or unconnected. I would hypothesize that during the creative process the brain begins by disorganizing, making links between shadowy forms of objects or symbols or words or remembered experiences that have not previously been linked. Out of this disorganization, self-organization eventually emerges and takes over in the brain. The result is a completely new and original thing: a mathematical function, a symphony, or a poem.[29]

5

INFINITE REVIEW: REMEMBERING AND FORGETTING

Pythagoras, the so-called "father of numbers," was said to have had a memory that surpassed any of his peers, extending back through a roll call of former lives that included Aethalides son of Hermes, Pyrrhus the fisherman, and Euphorbus the warrior. Belief in reincarnation—which had, Pythagoras estimated, occurred to him at 216-year intervals—was probably not widely shared outside Egypt and the East, and the possibility of past-life recollection was flatly rejected by early Christian thought. According to Saint Augustine, the most influential of the Church Fathers, Pythagoras had been the victim of "untrue recollections," most probably planted by malignant spirits.[1] In short, the great philosopher and mathematician's memory had fooled him.

More than any other aspect of human psychology, our understanding of the reach and function of memory has always been shaped by larger social issues and preoccupations. The rise of literate culture in Ancient Greece, for example, sparked concerns that everyday powers of recall might be damaged by a reliance on the written word. The alphabet would, Socrates warned, ultimately "create forgetfulness in the learners' souls," their memories undone by repeated outsourcing to the written word.[2]

There soon emerged a branch of scholarship that spoke directly to these concerns: the *ars memoriae*, which steered orators, statesmen, and students towards practical techniques of memory enforcement through association. Mnemonics remained a popular scholastic pursuit until the late Renaissance, yet its focus on memory as a form of conscious encodement and retrieval—or what today's psychologists would call *explicit memory*—tended to obscure the most vital and dynamic aspects of personal memory. Firstly, recollection was the binding thread of the self, the foundation of our personality, our way of being in the world. It was only through the persistence of our memories that, as John Locke observed, we remained the same person today as we were yesterday. Secondly, memory acts were for the most part unconscious, taking place without effort or awareness. To ride a bike, or to find oneself unexpectedly transported back to childhood by the taste of cake dipped in tea—both forms of memory involve chains of recall that cannot be easily accessed or verbalized.

Much of what modern science would go on to learn about the vagaries of personal memory would come from the study of its failures and limitations. Amnesia, first classified as a mental disorder in the mid-eighteenth century, revealed how older memories could often be fully retained by patients, while recall for recent events might be almost entirely obliterated. Moreover, while the behavior of an amnesiac patient sometimes betrayed knowledge of recent events for which they had no conscious knowledge, those with catastrophic memory loss showed a tendency, when questioned by clinicians, to confabulate, unthinkingly filling the gaps in their autographical memory with proxy recollections.

Most clinical cases of amnesia were organic in nature, largely precipitated by damage within, or near to, the hippocampus (the small, curved organ that sits beneath the cortex in the brain's temporal lobe) but by the late nineteenth century, cases of psychogenic or functional amnesia began to fall under the clinical spotlight, as neurologists such as Jean-Marie Charcot investigated how memory loss in hysterics might stem from emotional trauma. And it was this field of research that informed Sigmund Freud's first incursions into the mechanics of personal memory, leading him to postulate that repression was in fact a commonplace form of ego defense, its psychic blanketing of inner conflicts by no means limited to hysteria or neurosis.

Writers and philosophers before Freud had, of course, noted how our memories of trauma, wrongdoing, and distress were liable to become thoroughly clouded, as if to lessen their emotional pain and moral charge. Freud went further still. The strategic forgetfulness that was characteristic of hysteria was, he believed, also the director of everyday acts of absent-mindedness; it was the reason behind our limited recall for dreams and for the dark veil that made it all but impossible to remember anything of our infant lives. For Freud, forgetting was rarely a simple or innocent error, and it was never total. In contrast to pioneering psychologists such as Hermann Ebbinghaus, who proposed that the decay of memory followed a predictable "forgetting curve," Freud was convinced that "everything survives in one way or another, and is capable under certain circumstances of being brought back to life."[3]

New vistas in memory research began to open in Freud's own lifetime. Frederic Bartlett, Cambridge's first Professor of Experimental Psychology, succeeded in demonstrating the *constructive* nature of memory, using simple recollection tasks to show how memories were always dynamically refashioned. Barbara Milner's post-operative studies of memory impairments in epileptic patients (following the unilateral removal of the medial structures of the left temporal lobe) at the neurosurgery department in Hartford Hospital, Connecticut, revealed that long-term memory could be both declarative (explicit) and non-declarative (implicit). And research undertaken by various cognitive psychologists confirmed that the effective retention of everyday information required fast passage out of working memory: if not rehearsed within 15 to 30 seconds, a transient recollection would invariably decay and disappear.

But did memories ever really vanish? And, if not, could they be artificially revived? When the Canadian neurologist Wilder Penfield used electrical brain stimulation to stimulate flashbacks in his patients at Montreal Neurological Institute in the 1950s, the notion of a "memory cortex" containing a pure, ganglionic record of all our experiences was briefly and powerfully rekindled. Freudians nodded; scientologists went in search of deep-stored "engrams"; left-field psychiatrists turned to drugs such as LSD as memory enhancers, shepherding patients through perinatal memories and ancestral memories from the deep well of the "collective unconscious." Yet it was the fallibility of memory, its receptiveness to social cues and demand characteristics, that continued to fascinate experimental psychologists. False memories could, it was repeatedly found, quite easily be created in experimental subjects, particularly when seeded or corroborated by a "reliable" source. Under the right circumstances, most people could, like the great Pythagoras, be very easily led to recall events that had never happened.

*

For Plato (427–348 BC), recollection and memory were very different
processes. While recollection permitted access to universal and implicit truths,
memories were sensory records of everyday experience.

PLATO, *THEAETETUS* (CA. 369 BC)

Well then, let me ask you to suppose, for the sake of argument, that there's
an imprint-receiving piece of wax in our minds: bigger in some, smaller
in others; of cleaner wax in some, of dirtier in others; of harder wax in
some, of softer in others, but in some made of wax of a proper consistency. . . . And let's say it's the gift of Memory, the mother of the Muses;
and that if there's anything we want to remember, among the things we
see, hear, or ourselves conceive, we hold it under the perceptions and
conceptions and imprint them on it, as if we were taking the impressions
of signet rings. Whatever is imprinted, we remember and know, as long
as its image is present; but whatever is smudged out or proves unable to
be imprinted, we've forgotten and don't know.[4]

Aristotle (384–322 BC) grasped that memories were not only "imprinted" but
connected by associations that were essential to their recollection.

ARISTOTLE, *ON MEMORY* (350 BC)

It often happens that, though a person cannot recollect at the moment,
yet by seeking he can do so, and discovers what he seeks. This he suc-
ceeds in doing by setting up many movements, until finally he excites
one of a kind which will have for its sequel the fact he wishes to recollect.
For remembering is the existence of a movement capable of stimulating
the mind to the desired movement, and this, as has been said, in such a
way that the person should be moved from within himself, i.e., in consequence of movements wholly contained within himself. . . .

But one must get hold of a starting-point. This explains why it is that
persons are supposed to recollect sometimes by starting from 'places'. The
cause is that they pass swiftly from one point to another, e.g. from milk to
white, from white to mist, and thence to moist, from which one remem-
bers Autumn if this be the season he is trying to recollect. . . .

The cause of one's sometimes recollecting and sometimes not, though starting from the same point, is, that from the same starting-point a movement can be made in several directions, as, for instance, from C to B or to D. If, then, the mind has not moved in an old path, it tends to move to the more customary; for custom now assumes the role of nature. Hence the rapidity with which we recollect what we frequently think about. For as one thing follows another by nature, so too that happens by custom; and frequency creates nature. And since in the realm of nature occurrences take place which are even contrary to nature, or fortuitous, the same happens a fortiori in the sphere swayed by custom, since in this sphere nature is not similarly established. Hence it is that the mind receives an impulse to move sometimes in the required direction, and at other times otherwise, particularly when something else somehow deflects the mind from the right direction and attracts it to itself. This last consideration explains too how it happens that, when we want to remember a name, if we know one somewhat like it, we blunder on to that. Thus, then, recollection takes place.[5]

The Roman lawyer and scholar Marcus Tullius Cicero (106–43 BC),
considered the greatest orator of his time, relied heavily on an "artificial" aid
to memory consolidation.

CICERO, *DE ORATORE* (55 BC)

I am grateful to Simonides of Ceos, the reputed originator of the system of artificial memory. It is related that on one occasion, when he was supping with Scopas at Crannon, in Thessaly, and engaged in reciting some verses which he had composed in honour of that very prosperous and noble personage, he introduced, by way of embellishment, much poetical allusion to Castor and Pollux. At the conclusion, Scopas told him, in rather too sordid a spirit, that only half the stipulated sum should be paid him for his poem, for the other moiety, he might look, if he chose, to the Tyndaridae, who had engrossed full half of the eulogy. Shortly after, a message was said to have been brought to Simonides, that he was wanted at the door, where two young men were eagerly inquiring for him; he immediately rose and went out, but saw nobody. In the short interval of his absence, however, the hall where Scopas was banqueting with his friends fell in, crushing him and the whole party to death, and burying them in the ruins. When the mangled remains could not by any means be identified by their friends, who came to recover the bodies, Simonides had so distinct a recollection of the exact spot occupied by each individual that he was able to give satisfactory directions for their interment. Taking a hint from this occurrence, he is said to have discovered that order was the luminous guide to memory, and that those, therefore, who wish to cultivate this faculty should have places portioned off in the mind, fixing in these several compartments certain images to represent the ideas they wished to remember; thus the order of places would preserve the order of ideas, and the symbols would suggest the ideas themselves—the places standing for the wax and the images for the letters. . . .

But a verbal memory, which is not so necessary for us, must be distinguished by a greater variety of symbols: there are many words connecting, like joints, the different members of language which cannot be represented by any corresponding images; for these certain arbitrary symbols must be invented to be always used in their stead. But the memory of things is properly the memory of the orator, and this we may attain by the creation of distinct and aptly arranged images, so that the signs shall suggest the sentences and the compartments their regular succession. Nor is there any foundation for the objection of the indolent, that the memory is likely to be oppressed with the load of images, and that ideas easily retained by the natural memory are only rendered the more obscure by this artificial process; for I remember having seen two remarkable men with memories of almost superhuman tenacity, viz., Charmadas at Athens, and Scepsius Metrodorus in Asia, the latter of whom is said to be still living, by each of whom I was assured that he could inscribe in the different compartments of his mind whatever he wished to remember as easily as he could trace the letters in wax. Though memory, therefore, cannot be wrought out of the mind unless implanted there by nature, if latent it certainly may be elicited.[6]

John Locke (1632–1704) was one of the first philosophers to link personal identity to the continuity of memory. What made someone the "same person" over time, he suggested, is not the substance of the soul or body but rather

the continuity of consciousness, especially the ability to remember past
experiences.

JOHN LOCKE, *AN ESSAY CONCERNING HUMAN UNDERSTANDING* (1690)

The memory in some men, 'tis true, is very tenacious, even to a miracle: but yet there seems to be a constant decay of all our ideas, even of those which are struck deepest, and in minds the most retentive; so that if they be not sometimes renewed by repeated exercise of the senses, or reflection on those kind of objects, which at first occasioned them, the print wears out, and at last there remains nothing to be seen. Thus the ideas, as well as children, of our youth, often die before us: and our minds represent to us those tombs, to which we are approaching; where though the brass and marble remain, yet the inscriptions are effaced by time, and the imagery moulders away. *The pictures drawn in our minds, are laid in fading colours*; and if not sometimes refreshed, vanish and disappear. How much the constitution of our bodies, and the make of our animal spirits, are concerned in this, and whether the temper of the brain make this difference, that in some it retains the characters drawn on it like marble, in others like freestone, and in others little better than sand, I shall not here inquire, though it may seem probable, that the constitution of the body does sometimes influence the memory; since we oftentimes find a disease quite strip the mind of all its ideas, and the flames of a fever, in a few days, calcine all those images to dust and confusion, which seemed to be as lasting, as if graved in marble.[7]

Blessed with a prodigious memory, the writer and lexicographer Samuel Johnson (1709–1784) grasped that forgetfulness could be a psychological purgative, ridding the mind of recrimination and self-doubt.

SAMUEL JOHNSON, "THE REGULATION OF MEMORY" (1759)

It would add much to human happiness, if an art could be taught of forgetting all of which the remembrance is at once useless and afflictive, if that pain which never can end in pleasure could be driven totally away, that the mind might perform its functions without incumbrance, and the past might no longer encroach upon the present.

Little can be done well to which the whole mind is not applied; the business of every day calls for the day to which it is assigned, and he will have little leisure to regret yesterday's vexations who resolves not to have a new subject of regret to-morrow.

But to forget or to remember at pleasure, are equally beyond the power of man. Yet as memory may be assisted by method, and the decays of knowledge repaired by stated times of recollection, so the power of forgetting is capable of improvement. Reason will, by a resolute contest, prevail over imagination, and the power may be obtained of transferring the attention as judgment shall direct.

The incursions of troublesome thoughts are often violent and importunate; and it is not easy to a mind accustomed to their inroads to expel them immediately by putting better images into motion; but this enemy of quiet is above all others weakened by every defeat; the reflection which has been once overpowered and ejected, seldom returns with any formidable power.

Employment is the great instrument of intellectual dominion. The mind cannot retire from its enemy into total vacancy, or turn aside from one object but by passing to another. The gloomy and the resentful are always found among those who have nothing to do, or that do nothing. We must be busy about good or evil, and he to whom the present offers nothing will be looking backward on the past.[8]

The novelist Thomas Mann believed that Freud's theory of repression was taken wholesale from Arthur Schopenhauer (1788–1860) and "translated from metaphysics into psychology."

ARTHUR SCHOPENHAUER, *WORLD AS WILL AND REPRESENTATION* (1818)

[R]emember how reluctantly we think of things that powerfully prejudice our interests, wound our pride, or interfere with our wishes; with what difficulty we decide to lay such things before our own intellect for accurate and serious investigation; how easily, on the other hand, we unconsciously break away or sneak off from them again; how, on the contrary, pleasant affairs come into our minds entirely of their own accord, and, if driven away, always creep on us once more, so that we dwell on them for hours. In this resistance on the part of the will to allow what is contrary to it to come under the examination of the intellect is to be found the place where madness can break in on the mind. Every new adverse event must be assimilated by the intellect, in other words, must receive a place in the system of truths connected with our will and its interests, whatever it may have to displace that is more satisfactory. As soon as this is done, it pains us much less; but this operation itself is often very painful, and in most cases takes place only slowly and with reluctance. But soundness of mind can continue only in so far as this operation has been correctly carried out each time. On the other hand, if, in a particular case, the resistance and opposition of the will to the assimilation of some knowledge reaches such a degree that that operation is not clearly carried through; accordingly, if certain events or circumstances are wholly suppressed for the intellect, because the will cannot bear the sight of them; and then, if the resultant gaps are arbitrarily filled up for the sake of the necessary connexion; we then have madness. For the intellect has given up its nature to please the will; the person then imagines what does not exist. But the resultant madness then becomes the Lethe of unbearable sufferings; it was the last resource of worried and tormented nature, i.e., of the will.[9]

The Scottish churchman and economist Thomas Chalmers (1780–1847) was one of the few nineteenth-century intellectuals to consider memory an inherently social and not merely personal resource.

THOMAS CHALMERS, *THE ADAPTATION OF EXTERNAL NATURE TO THE MORAL AND INTELLECTUAL CONSTITUTION OF MAN* (1835)

Let us imagine for example, that a daily companion had, unknown to us, kept a minute and statistical journal of all the events we personally shared in; and the likelihood is, that, if admitted to the perusal of this document, even after the lapse of half a life time, our memory would depone to many thousand events which had else escaped, into utter and irrecoverable forgetfulness. It is certainly remarkable, that, on some brief utterance by another, the stories of former days should suddenly reappear, as if in illumined characters, on the tablet from which they had so totally faded; that the mention of a single circumstance, if only the link of a train, should conjure to life again a whole host of sleeping recollections: And so, in each of our fellow-men, might we have a remembrancer, who can vivify our consciousness anew, respecting scenes and transactions of our former history which had long gone by; and which, after having vanished once from a solitary mind left to its own processes, would have vanished everlastingly. . . .

It is thus, that, not only can one man make instant translation of his own memory; but on certain subjects, he can even make instant translation of his own intelligence into the mind of another. A shrewd discerner of the heart, when laying open its heretofore unrevealed mysteries, makes mention of things which at the moment we feel to be novelties; but which, almost at the same moment, are felt and recognised by us as truths—and that, not because we receive them upon his authority, but on the independent view that ourselves have of their own evidence. His utterance, in fact, has evoked from the cell of their imprisonment, remembrances, which but for him, might never have been awakened.[10]

Thomas De Quincey's (1785–1859) most searching ruminations on opium's power of reviving memory appeared in short essays published in Blackwood's Magazine.

THOMAS DE QUINCEY, *SUSPIRIA DE PROFUNDIS* (1845)

Yes, reader, countless are the mysterious hand-writings of grief or joy which have inscribed themselves successively upon the palimpsest of your brain; and, like the annual leaves of aboriginal forests, or the undissolving snows on the Himalaya, or light falling upon light, the endless strata have covered up each other in forgetfulness. But by the hour of death, but by fever, but by the searchings of opium, all these can revive in strength. They are not dead, but sleeping. In the illustration imagined by myself, from the case of some individual palimpsest, the Grecian tragedy had seemed to be displaced, but was not displaced, by the monkish legend; and the monkish legend had seemed to be displaced, but was not displaced, by the knightly romance. In some potent convulsion of the system, all wheels back into its earliest elementary stage. The bewildering romance, light tarnished with darkness, the semi-fabulous legend, truth celestial mixed with human falsehoods, these fade even of themselves, as life advances. The romance has perished that the young man adored; the legend has gone that deluded the boy; but the deep, deep tragedies of infancy, as when the child's hands were unlinked forever from his mother's neck, or his lips forever from his sister's kisses, these remain lurking below all, and these lurk to the last. Alchemy there is none of passion or disease that can scorch away these immortal impresses; and the dream which closed the preceding section, together with the succeeding dreams of this . . . are but illustrations of this truth, such as every man probably will meet experimentally who passes through similar convulsions of dreaming or delirium from any similar or equal disturbance in his nature.[11]

In his seminal study of memory deficits and abnormalities, the French psychologist Théodule-Armand Ribot (1839–1916) proposed the so-called Law of Regression.

THÉODULE-ARMAND RIBOT, *DISEASES OF MEMORY* (1882)

We have stated these two facts in the dissolution of memory: the new perishes before the old, the complex before the simple. The law which we have formulated is only the psychological expression of a law of life, and pathology shows in its turn that memory is a biological fact. The study of periodic amnesia has thrown much light upon our subject. In showing us how the memory is dissolved and reconstructed, it teaches us what memory is. It has revealed a law which permits us to observe morbid types in great variety and from many points of view; later on, we shall, by its aid, be able to include them in one general survey.

Without attempting a careful review in this place, let us recall briefly what has been observed above. First, in all cases, abolition of recent impressions; in periodic amnesia, total suspension of all forms of memory,

except those which are semi-organized and organic; in total and temporary amnesia, complete loss of memory, except in its organic forms; in
one instance (Macnish) amnesia comprising its organic forms. We shall
see in the following that partial disorders of memory are governed by this
same law of regression, and especially the most important group—that of
amnesia of language.

The law of regression being admitted, we have now to determine in
what matter it acts. Upon this point I shall be brief, having only hypotheses to offer. It would be puerile to suppose that recollections are arranged
in the brain in the form of layers in order of age, after the fashion of
geological strata, and that disease, penetrating from the surface to the
lowest point, acts like an experimentalist removing the brain of an animal, bit by bit. To explain the action of the morbid process we must have
recourse to the hypothesis advanced above the regard to the physical
bases of memory. It may be summed up in a few words.

It is very probable that recollections occupy the same anatomical seat
as primitive impressions, and that they excite the activity of the same
nervous elements (cells and filaments). The latter may have very different
positions from the surface of the brain to the spinal cord. Conservation
and reproduction depend: (1) upon a certain modification of the cells; (2)
upon the formation of more or less complex groups which we have des-
ignated as dynamic associations. Such are the physical bases of memory.

Primitive acquisitions, those that date from infancy, are the most simple; they include the formation of secondary automatic movements in
the education of the senses. They depend principally upon the medulla
and the lower centers of the brain; and we know that at this period of life
the exterior cerebral layers are imperfectly developed. Aside from their
simplicity there is every reason why these acquisitions should be stable.
In the first place, the impressions are received in virgin elements. Nutri-
tion is very active; but incessant molecular repair serves only to fix the
registered perception; the new molecules taking the exact places occupied by the old, the acquired state finally becomes organic. Moreover, the
dynamic associations formed between the different elements attain after
a time to a condition of complete fusion, thanks to continual repetition.
It is inevitable, then, that the earlier acquisitions be better conserved and

more easily reproduced than any others, and that they should constitute the most lasting form of memory.[12]

Some of Sigmund Freud's (1856–1939) earliest thoughts on the dynamic and repressive functions of memory and repression appeared in his letters to Wilhelm Fliess.

SIGMUND FREUD, LETTER TO WILHELM FLIESS (DECEMBER 6, 1896)

As you know, I am working on the assumption that our psychic mechanism has come into being by a process of stratification: the material present in the form of memory traces being subjected from time to time to a *rearrangement* in accordance with fresh circumstances—to a *retranscription*. . . .

I should like to emphasize the fact that the successive registrations represent the psychic achievement of successive epochs of life. At the boundary between two such epochs a translation of the psychic material must take place. I explain the peculiarities of the psychoneuroses by supposing that this translation has not taken place in the case of some of the material, which has certain consequences. For we hold firmly to a belief in a tendency toward quantitative adjustment. Every later transcript inhibits its predecessor and drains the excitatory process from it. If a later transcript is lacking, the excitation is dealt with in accordance with the psychological laws in force in the earlier psychic period and along the paths open at that time. Thus an anachronism persists: in a particular province, *"fueros"* are still in force; we are in the presence of "survivals." A failure of translation—this is what is known clinically as "repression." The motive for it is always a release of the unpleasure that would be generated by a translation; it is as though this unpleasure provokes a disturbance of thought that does not permit the work of translation. Within one and the same psychic phase, and among registration of the same kind, a normal defense makes itself felt owing to a generation of unpleasure. But pathological defense occurs only against a memory trace from an earlier phase that has not yet been translated.[13]

Henri Bergson's (1859–1941) great insight was to recognize the existence of different memory systems.

HENRI BERGSON, *MATTER AND MEMORY* (1896)

[W]e are confronted by two different memories theoretically independent. The first records, in the form of memory-images, all the events of our daily life as they occur in time; it neglects no detail; it leaves to each fact, to each gesture, its place and date. Regardless of utility or of practical application, it stores up the past by the mere necessity of its own nature. By this memory is made possible the intelligent, or rather intellectual, recognition of a perception already experienced; in it we take refuge every time that, in the search for a particular image, we remount the slope of our past. But every perception is prolonged into a nascent action; and while the images are taking their place and order in this memory, the movements which continue them modify the organism, and create in the body new dispositions towards action. Thus is gradually formed an experience of an entirely different order, which accumulates within the body, a series of mechanisms wound up and ready, with reactions to external stimuli ever more numerous and more varied, and answers ready prepared to an ever growing number of possible solicitations. We become conscious of these mechanisms as they come into play; and this consciousness of a whole past of efforts stored up in the present is indeed also a memory, but a memory profoundly different from the first, always bent upon action, seated in the present and looking only to the future. It has retained from the past only the intelligently coordinated movements which represent the accumulated efforts of the past; and it recovers those past efforts, not in the memory-images which recall them, but in the definite order and systematic character with which the actual movements take place. In truth, it no longer *represents* our past to us, it *acts* it; and if it still deserves the name of memory, it is not because it conserves bygone images, but because it prolongs their useful effect into the present moment.[14]

The Swiss psychiatrist Theodore Flournoy (1854–1920) believed that all supernatural experiences could be ultimately explained in terms of psychology. In 1900, Flournoy coined the term cryptomnesia to describe the subconscious interplay of memory and imagination in the séances of Geneva medium Hélène Smith, whose trance visitations displayed intimate knowledge of Martian language and several previous incarnations.

THEODOR FLOURNOY, *FROM INDIA TO THE PLANET MARS* (1900)

While the Martian romance is purely a work of fantasy, in which the creative imagination was able to allow itself free play through having no investigation to fear, the Hindoo cycle, and that of Marie Antoinette, having a fixed terrestrial setting, represent a labor of construction which was subjected from the start to very complex conditions of environments and epochs. To keep within the bounds of probability, not to be guilty of too many anachronisms, to satisfy the multiple demands of both logic and æsthetics, formed a particularly dangerous undertaking, and one apparently altogether beyond the powers of a person without special instruction in such matters. The subconscious genius of Mlle. Smith has acquitted itself of the task in a remarkable manner, and has displayed in it a truly wonderful and delicate sense of historic possibilities and of local color.

The Hindoo romance, in particular, remains for those who have taken part in it a psychological enigma, not yet solved in a satisfactory manner, because it reveals and implies in regard to Hélène, a knowledge relative to the costumes and languages of the Orient, the actual source of which it has up to the present time not been possible to discover. All the witnesses of Mlle. Smith's Hindoo somnambulisms who are of the same opinion on that subject (several refrain from having any) unite in seeing in it a curious phenomenon of cryptomnesia, of reappearances of memories profoundly buried beneath the normal waking state, together with an indeterminate amount of imaginative exaggeration upon the canvas of actual facts. But by this name of cryptomnesia, or resurrection of latent memories, two singularly different things are understood. For me it is only a question of memories of her present life; and I see nothing of the supernormal in that. For while I have not yet succeeded in finding the key to the enigma, I do not doubt its existence, and I will mention later certain indications which seem to me to support my idea that the Asiatic notions of Mlle. Smith have a wholly natural origin.[15]

*The Swiss psychologist Édouard Claparède (1873–1940) found that
amnesiacs could retain memories of recent events that they could not
directly access.*

ÉDOUARD CLAPARÈDE, "RECOGNITION AND 'ME-NESS'" (1911)

The patient was a woman hospitalized at Asile de Bel-Air. She was 47 at
the time of the first experiment, 1906. Her illness had started around
1900. Her old memories remained intact: she could correctly name the
capitals of Europe, make mental calculations, and so on. But she did not
know where she was, though she had been at the asylum five years. She
did not recognize the doctors whom she saw every day, nor her nurse
who had been with her for six months. When the latter asked the patient
whether she knew her, the patient said: "No, Madame, with whom have
I the honor of speaking?" She forgot from one minute to the next what
she was told, or the events that took place. She did not know what year,
month, and day it was, though she was being told constantly. She did not
know her age, but could figure it out if told the date.

I was able to show, by means of learning experiments done by the
method, that not all ability of mnemonic registration was lost in this person. What is worthy of our attention here was her inability to evoke recent
memories voluntarily, while they did arise automatically, by chance, as
recognitions.

When one told her a little story, read to her various items of a newspa-
per, three minutes later she remembered nothing, not even the fact that
someone had read to her; but with certain questions one could elicit in
a reflex fashion some of the details of those items. But when she found
these details in her consciousness, she did not recognize them as memories but believed them to be something "that went through her mind" by
chance, an idea she had "without knowing why," a product of her imagination of the moment, or even the result of reflection.

I carried out the following curious experiment on her: to see whether
she would better retain an intense impression involving affectivity, I
stuck her hand with a pin hidden between my fingers. The light pain
was as quickly forgotten as indifferent perceptions; a few minutes later

she no longer remembered it. But when I again reached out for her hand, she pulled it back in a reflex fashion, not knowing why. When I asked for the reason, she said in a flurry, "Doesn't one have the right to withdraw her hand?" and when I insisted, she said, "Is there perhaps a pin hidden in your hand?" To the question, "What makes you suspect me of wanting to stick you?" she would repeat her old statement, "That was an idea that went through my mind," or she would explain, "Sometimes pins are hidden in people's hands." But never would she recognize the idea of sticking as a "memory."[16]

The Cambridge psychologist Frederic Bartlett (1886–1969) experimentally demonstrated how memories came to life through "interests" that determined their shape and meaning.

FREDERIC BARTLETT, *REMEMBERING* (1932)

Remembering is not the re-excitation of innumerable fixed, lifeless and fragmentary traces. It is an imaginative reconstruction, or construction, built out of the relation of our attitude towards a whole active mass of organised past reactions or experience, and to a little outstanding detail which commonly appears in image or in language form. It is thus hardly ever really exact, even in the most rudimentary cases of rote recapitulation, and it is not at all important that it should be so. The attitude is literally an effect of the organism's capacity to turn round upon its own 'schemata', and is directly a function of consciousness. The outstanding detail is the result of that valuation of items in an organised mass which begins with the functioning of appetite and instinct, and goes much further with the growth of interests and ideals. Even apart from their appearance in the form of sensorial images, or as language forms, some of the items of a mass may stand out by virtue of their possession of certain physical characteristics. But there is no evidence that these can operate in determining a specific reaction, except after relatively short periods of delay. The active settings which are chiefly important at the level of human remembering are mainly 'interest' settings; and, since an interest has both a definite direction and a wide range, the development of these settings involves much reorganisation of the 'schemata' that follow the more primitive lines of special sense differences, of appetite and of instinct. So, since many 'schemata' are built of common materials, the images and words that mark some of their salient features are in constant, but explicable, change. They, too, are a device made possible by the appearance, or discovery, of consciousness, and without them no genuine long-distance remembering would be possible.[17]

Beyond the realm of personal memory, there was, according to the Swiss psychologist Carl Gustav Jung (1875–1961), a deep and universal well of ancestral memory.

CARL GUSTAV JUNG, *THE CONCEPT OF THE COLLECTIVE UNCONSCIOUS* (1936)

The collective unconscious is a part of the psyche which can be negatively distinguished from a personal unconscious by the fact that it does not, like the latter, owe its existence to personal experience and consequently is not a personal acquisition. While the personal unconscious is made up essentially of contents which have at one time been conscious but which have disappeared from consciousness through having been forgotten or repressed, the contents of the collective unconscious have never been in consciousness, and therefore have never been individually acquired, but owe their existence exclusively to heredity. Whereas the personal unconscious consists for the most part of complexes, the content of the collective unconscious is made up essentially of archetypes.

The concept of the archetype, which is an indispensable correlate of the idea of the collective unconscious, indicates the existence of definite forms in the psyche which seem to be present always and everywhere. Mythological research calls them "motifs"; in the psychology of primitives they correspond to Lévy-Bruhl's concept of "représentations collectives," and in the field of comparative religion they have been defined by Hubert and Mauss as "categories of the imagination." Adolf Bastian long ago called them "elementary" or "primordial thoughts." From these references it should be clear enough that my idea of the archetype—literally a pre-existent form—does not stand alone but is something that is recognized and named in other fields of knowledge.

My thesis, then, is as follows: In addition to our immediate consciousness, which is of a thoroughly personal nature and which we believe to be the only empirical psyche (even if we tack on the personal unconscious as an appendix), there exists a second psychic system of a collective, universal, and impersonal nature which is identical in all individuals. This collective unconscious does not develop individually but is inherited. It consists of pre-existent forms, the archetypes, which can only become conscious secondarily and which give definite form to certain psychic contents.[18]

While Freud leant heavily on Darwinian theory in both his psychoanalytic and sociological writings, he remained, like Jung, wedded to a Lamarckian notion of hereditary transmission via memory-traces.

SIGMUND FREUD, *MOSES AND MONOTHEISM* (1939)

[A]llow me to venture further and assert that the archaic heritage of mankind includes not only dispositions, but also ideational contents, memory-traces of the experiences of former generations. In this way the extent as well as the significance of the archaic heritage would be enhanced in a remarkable degree.

On second thoughts I must admit that I have argued as if there were no question that there exists an inheritance of memory-traces of what our forefathers experienced, quite independently of direct communication and of the influence of education by example. When I speak of an old tradition still alive in a people, of the formation of a national character, it is such an inherited tradition and not one carried on by word of mouth that I have in mind. Or at least I did not distinguish between the two, and was not quite clear about what a bold step I took by neglecting this difference. This state of affairs is made more difficult, it is true, by the present attitude of biological science which rejects the idea of acquired qualities being transmitted to descendants. I admit, in all modesty, that in spite of this I cannot picture biological development proceeding without taking this factor into account.[19]

In the ragtag pseudoscience of Dianetics, all psychosomatic illness was caused by blocked "engrams" which the trained "auditor" was assigned to edit and refile.

L. RON HUBBARD, *DIANETICS: THE MODERN SCIENCE OF MENTAL HEALTH* (1950)

The auditor, with precision methods, recovers data from the earliest "unconscious" moments of the patient's life, such "unconsciousness" being understood to be caused by shock or pain, not mere unawareness. The patient thus contacts the cellular-level engrams. Returned to them and progressed through them by the auditor, the patient re-experiences these moments a few times, when they are then erased and refiled automatically as standard memory. So far as the auditor and the patient can discover, the entire incident has now vanished and does not exist. If they searched carefully in the standard banks, they would find it again, but refiled as "Once aberrative, do not permit as such into computer." Late areas of "unconsciousness" are impenetrable until early ones are erased. The amount of discomfort experienced by the patient is minor. He is repelled mainly by engramic commands which variously dictate emotion and reaction. In a Release, the case is not progressed to the point of complete recall. In a Clear, full memory exists throughout the lifetime, with the additional bonus that he has photographic recall in color, motion, sound, etc., as well as optimum computational ability. The psychosomatic illnesses of the Release are reduced, ordinarily, to a level where they do not thereafter trouble him. In a Clear, psychosomatic illness has become non-existent and will not return since its actual source is nullified permanently.[20]

The renowned Russian neurologist Alexander Luria (1902–1977) began studying the formidable memory of Moscow journalist Solomon Shereshevsky in the 1920s. Shereshevsky was a synesthete whose multi-sensory relationship to language underscored an apparently limitless memory.

ALEXANDER LURIA, *THE MIND OF MNEMONIST* (1968)

Yet how was one to explain forgetting in a man whose memory seemed inexhaustible? How [to] explain that sometimes there were instances in which S. *omitted* some elements in his recall but scarcely ever *reproduced material inaccurately* (by substituting a synonym or a word closely associated in meaning with the one he'd been given)?

The experiments immediately turned up answers to both questions. S. did not "forget" words he'd been given; what happened was that he omitted these as he "read off" a series. And in each case there was a simple explanation for the omissions. If S. had placed a particular image in a spot where it would be difficult for him to "discern"—if he, for example, had placed it in an area that was poorly lit or in a spot where he would have trouble distinguishing the object from the background against which it had been set—he would omit this image when he "read off" the series he had distributed along his mental route. He would simply walk on "without noticing" the particular item, as he explained.

These omissions (and they were quite frequent in the early period of our observation, when S.'s technique of recall had not developed to its fullest) clearly were not defects of memory but were, in fact, defects of perception. They could not be explained in terms of established ideas on the neurodynamics of memory traces (retroactive and proactive inhibition, extinction of traces, etc.) but rather by certain factors that influence perception (clarity, contrast, the ability to isolate a figure from its background, the degree of lighting available, etc.). His errors could not be explained, then, in terms of the psychology of memory but had to do with the psychological factors that govern perception.[21]

In the late 1930s, the pioneering neurosurgeon Wilder Penfield (1891–1976) found that stimulation of specific areas of the temporal lobe caused some patients at the Montreal Neurological Institute Hospital to report vivid "flashbacks" of seemingly forgotten memories.

WILDER PENFIELD, *MYSTERY OF THE MIND* (1975)

In the course of surgical treatment of patients suffering from temporal lobe seizures (epileptic seizures that are caused by a discharge that originates in that lobe), we stumbled upon the fact that electrical stimulation of the interpretive areas of the cortex occasionally produces what Hughlings Jackson had called "dreamy states," or "psychical seizures." Sometimes the patient informed us that we had produced one of his "dreamy states" and we accepted this as evidence that we were close to the cause of his seizures. It was evident at once that these were not dreams. They were electrical activations of the sequential record of consciousness, a record that had been laid down during the patient's earlier experience. The patient "re-lived" all that he had been aware of in that earlier period of time as in a moving-picture "flashback."

On the first occasion, when one of these "flashbacks" was reported to me by a conscious patient (1933), I was incredulous. On each subsequent occasion, I marvelled. For example, when a mother told me she was suddenly aware, as my electrode touched the cortex, of being in her kitchen listening to the voice of her little boy who was playing outside in the yard. She was aware of the neighborhood noises, such as passing motor cars, that might mean danger to him.

A young man stated he was sitting at a baseball game in a small town and watching a little boy crawl under the fence to join the audience. Another was in a concert hall listening to music.[22]

The first clinical reports of confabulation emerged in the late 1880s, when the Russian psychiatrist Sergei Korsakoff studied the memory distortions of alcoholic patients with brain damage. The following case of anterograde amnesia, described by the psychologist Howard Gardner (1943–), underscored that confabulation was a largely unconscious process that attempted to compensate for missing memories.

HOWARD GARDNER, *THE SHATTERED MIND* (1977)

The patient was invariably cheerful, helpful and considerate. He would patiently execute tasks and respond to questions for hours on end, one major reason being that he had no recollection of ever tackling the task before. The only times in which he displayed even mild annoyance came when I pressed him on a question for which he had no ready answer. Even then, however, any trace of the anger evaporated upon the next question. As long as the conversation was restricted to small talk and pleasantries, Mr O'Donnell seemed entirely normal. And his knowledge of events of his childhood and the early years of his adulthood, of those spent in the Service, in college, and as a young entrepreneur, seemed generally intact. . . .

When asked questions he could not answer, he would give responses that were patently absurd. Questioned as to where he had seen me before, he immediately replied that we had gone fishing together for many years in Canada. Queried about the name of the hospital, he replied 'Why, this is the Quincy branch of the Massachusetts General Hospital.' A question about where he had been entering the hospital elicited the answer, 'Why, I'm just back from Korea, where I've been on active duty.'

On the surface, such responses seem to be total fabrications—confabulations in the pejorative sense. It is difficult to avoid suspicion that the patient is putting you on. Actually, however, there was no reason to think that Mr O'Donnell's answers were insincere. He was not lying, nor was he spinning a tall tale to amuse or confuse that examiner; rather, he was desperately trying to make momentary sense out of a bewildering situation. Accordingly, he latched on to any fragment of past reality that was accessible and that seemed to fit plausibly into the present context. O'Donnell *had*, in fact, as scattered points in his life, gone on fishing trips to Canada, visited Massachusetts General Hospital, rested (on his return from Korea) in an Army hospital. What most immediately prompted these anachronistic responses was the association of something in his present environment—my resemblance to a fishing companion, occasional pains in his leg, his Army serial number on his wrist band—to the thrust of a given question. The confabulation, then, represented to a legitimate effort to achieve some sense of coherence, some stable orientation that was otherwise, given his gross memory disturbances, unattainable.[23]

The cognitive scientist Marvin Minsky (1927–2016) considers why we have such limited recall of our early childhood.

MARVIN MINSKY, *SOCIETY OF MIND* (1986)

Ask anyone for memories from childhood, and everyone will readily produce a handful of stories like this:

My neighbor's father died when I was four. I remember sitting with my friend in front of their house, watching people come and go. It was strange. No one said anything.

It's hard to distinguish memories from memories of memories. Indeed, there's little evidence that any of our adult memories really go way back to infancy; what seem like early memories may be nothing more than reconstructions of our older thoughts. For one thing, recollections from our first five years seem strangely isolated; if we ask what happened earlier that day, the answer almost always is, "I don't remember that." Furthermore, many of those early memories involve incidents so significant that they probably occupied the child's mind repeatedly over a period of years. Most suspicious of all is the fact that such recollections are frequently described as seen through other, older eyes—with the narrator portrayed inside the scene, right near the center of the stage. Since we never actually see ourselves, these must be reconstructed memories, rehearsed and reformulated since infancy.

I suspect that this "amnesia of infancy" is no mere effect of decay over time but an inevitable result of growing out of infancy. A memory is not a separate entity, apart from how it works upon the mind. To remember an early experience, you must be able not only to "retrieve" some old records, but to reconstruct how your earlier mind reacted to them—and to do that, you would have to become an infant again. To outgrow infancy, you have to sacrifice your memories because they're written in an ancient script that your later selves can no longer read.[24]

The Estonian-born psychologist Endel Tulving (1927–2023) was a pioneer of experimental studies of encoding and retrieval. According to Tulving, "[n]o profound generalizations can be made about memory as a whole, but general statements about particular kinds of memory are perfectly possible."

ENDEL TULVING AND DANIEL L. SCHACTER, "PRIMING AND HUMAN MEMORY SYSTEMS" (1990)

Evidence is accumulating about yet another category of learning and memory, one that is not procedural, semantic or episodic. It has come to be known as priming. Its function is to improve identification of perceptual objects. Priming is a type of implicit memory; it does not involve explicit or conscious recollection of any previous experiences. It has affinities to both procedural memory in that it enhances perceptual skills. It also resembles semantic memory in that it involves cognitive representations of the world and expresses itself in cognition rather than behavior.

The prototypical priming experiment consists of two stages. In the first (study) stage, the subject is presented with a stimulus object (target). Target stimuli may comprise words, line drawings of objects, drawings of faces, and the like. In the second (test) stage, which may follow the first after an interval that can vary from seconds to months, the subject is given reduced perceptual information about the object and asked to name or categorize it. Reduced cues may consist of initial letters of graphemic fragments of words, partially obliterated words or figures, originally presented faces in a more highly schematized form, or tachistoscopic presentation of stimuli. Priming is said to have been demonstrated if the probability of the identification of the previously encountered targets is increased, or the latency of the identification response is reduced, in comparison with similar measures of nonstudied control items. The difference between performance on the target items and the nonstudied items provides a measure of the magnitude of the priming effect. . . .

We believe that priming and perceptual identification are expressions of a single perceptual representation system (PRS), which exists separately but interacts closely with other memory systems.[25]

In Elizabeth Loftus's (1944–) seminal experiment on recovered memory, adult subjects were presented with a booklet citing three events that had, according to relatives, occurred when they were children. The one false event described was a shopping trip, undertaken at the age of five, in which they had gone lost.

ELIZABETH LOFTUS, *CREATING FALSE MEMORIES* (1997)

The lost-in-the-mall scenario included the following elements: lost for an extended period, crying, aid and comfort by an elderly woman and, finally, reunion with the family. After reading each story in the booklet, the participants wrote what they remembered about the event. If they did not remember it, they were instructed to write, "I do not remember this." In two follow-up interviews, we told the participants that we were interested in examining how much detail they could remember and how their memories compared with those of their relative. The event

paragraphs were not read to them verbatim, but rather parts were provided as retrieval cues. The participants recalled something about 49 of
the 72 true events (68 percent) immediately after the initial reading of the
booklet and also in each of the two follow-up interviews. After reading
the booklet, seven of the 24 participants (29 percent) remembered either
partially or fully the false event constructed for them, and in the two
follow-up interviews six participants (25 percent) continued to claim that
they remembered the fictitious event. . . .

Research is beginning to give us an understanding of how false memories of complete, emotional and self-participatory experiences are created
in adults. First, there are social demands on individuals to remember;
for instance, researchers exert some pressure on participants in a study
to come up with memories. Second, memory construction by imagin-
ing events can be explicitly encouraged when people are having trouble
remembering. And, finally, individuals can be encouraged not to think
about whether their constructions are real or not. Creation of false
memories is most likely to occur when these external factors are present,
whether in an experimental setting, in a therapeutic setting or during
everyday activities.[26]

6

HOMO REPETITUS: FROM HABIT TO AUTOMATICITY

Towards the end of the eighteenth century, as Enlightenment freethinkers called time on the "vulgar errors" that were impeding the progress of science and rationality, the moral status of habit faced its first philosophical backlash. Habits were merely mechanical actions undertaken without will or thought, declared Thomas Reid in 1788, in *Essays on the Active Powers of Man*. If we were not mindful of their formation, we could all be slavishly "carried by habit as by a stream in swimming."[1]

Though Kant, Kierkegaard, and Nietzsche soon joined the rumbling counterblast against habitual thought and action, the need to carefully cultivate habits that would improve our intellect, morality, and productivity remained a subject upon which most educators, church men, and factory owners could agree. Habit's bundle of accidental and elected instincts, its propulsive power, confirmed to physiologists and psychologists that the brain and mind always worked to reduce the attentional load that any one task required.

Writing in 1890, William James was in no doubt that the "effortless custody" of habit was something that we should all cultivate, making as many useful acts as possible habitual and automatic: "Any sequence of mental action which has been frequently repeated tends to perpetuate itself; so that we find ourselves automatically prompted to think, feel, or do what we have been before accustomed to think, feel, or do, under like circumstances, without any consciously formed purpose, or anticipation of results."[2]

James' assertion that habits were "mechanically nothing but a reflex of discharge" would be largely confirmed by the work of the Russian physiologist Ivan Pavlov, whose research on salivation in dogs led him to find that *conditioned reflexes* could be created by pairing an inborn reflex with a neutral stimulus. Watch how any dog could be made to drool at the sound of a metronome, if the sound had previously and repeatedly attended the presentation of its food.

Though Pavlov himself stopped short of claiming that the principle of "classical conditioning" could be applied to "higher psychic activity in man," the brief was taken up with gusto by the American psychologist and founder of behaviorism John Broadus Watson. At the turn of the twentieth century, most psychological research on learning and reinforcement had concerned simple tasks such as typing and maze navigation, with researchers plotting learning curves that allowed them to measure the degree of acquired automaticity. With little to no experimental data of his own to draw on, Watson envisioned a Brave New World in which careful control of the environment could be used to direct an individual's entire repertoire of habits and abilities: "Give me a dozen healthy infants," Watson famously remarked, ". . . and my own specified world to bring them up in and I'll guarantee to take any one at random and train him to become any type of specialist I might select—doctor, lawyer, artist, merchant-chief, and, yes, even beggar man and thief, regardless of his talents, penchants, tendencies, abilities, vocations, and race of his ancestors."[3]

Watson's championing of the mind as *tabula rasa*, a blank page awaiting instructions for automation, put him at odds with psychologists and psychoanalysts who supposed that our thoughts and actions were offshoots of biological drives and unconscious instincts. But while Watson relished his role as an agent provocateur, behaviorism's failure to furnish any significant experimental data on the possibilities of manipulating human thought, feeling, and action hastened its demise. A decade after his overheated report on "Little Albert" (a nine-month-old infant who was conditioned to fear furry and hairy objects by coupling their exposure with the sound of a hammer), behaviorism hit the doldrums, its version of stimulus-response learning roundly dismissed as a white elephant. Writing in the *Atlantic Monthly* in 1934, one commentator went as far as accusing the fledgling science of having duped the public: "It has renamed our emotions 'complexes' and our habits 'conditioned reflexes,' but it has not changed our habits nor rid us of our emotions."[4]

The new science of habit formation sought to make good on Watson's promise over the coming decades, offering more nuanced theories of behavioral modification, as well as an unending series of therapies and interventions designed to break the cycle of addiction and dependency. Yet if expert and lay opinion continued to hold that all our actions and choices could be countermanded—or at the very least displaced through a replacement habit—findings from cognitive psychologists suggested otherwise.

According to the dual-process theory of mind developed by psychologists such as Keith Stanovich and Richard West, our response to environmental cues and information relies on two types of thinking: one fast, intuitive, and error-prone (Type 1); the other slow and deliberative (Type 2). The major characteristic of the "fast" skills repeatedly used in everyday judgements and decision making—from the ability to read simple sentences or recognize emotions in others—was that they could not be brought under attentional control.

Contrary to the behaviorist credo, for some mental shortcuts the "effortless custody" of automaticity is simply too intractable, too deeply fixed into our second nature, to be either modified or undone.

*

Closely tied to John Locke's notion of the mind as a tabula rasa was his theory of mental habit as the secret generator of our actions and judgements.

JOHN LOCKE, *AN ESSAY CONCERNING HUMAN UNDERSTANDING* (1690)

Habits, especially such as are begun very early, come at last to *produce actions in us, which often escape our observation.* How frequently do we, in a day, cover our eyes with our eyelids, without perceiving that we are at all in the dark! Men that, by custom, have got the use of a by-word, do almost in every sentence pronounce sounds which, though taken notice of by others, they themselves neither hear nor observe. And therefore it is not so strange, that our mind should often change the idea of its sensation into that of its judgement, and make one serve only to excite the other, without our taking notice of it.[5]

Following John Locke in denying the existence of innate ideas, David Hume (1711–1776) found that custom or habit was the "great guide of human life." In this respect, Hume's philosophy anticipates behavioral psychology and the weight that it gave to exposure and reinforcement in daily thought and action.

DAVID HUME, *AN ENQUIRY CONCERNING HUMAN UNDERSTANDING* (1748)

Suppose a person, though endowed with the strongest faculties of reason and reflection, to be brought on a sudden into this world; he would, indeed, immediately observe a continual succession of objects, and one event following another; but he would not be able to discover any thing farther. He would not, at first, by any reasoning, be able to reach the idea of cause and effect; since the particular powers, by which all natural operations are performed, never appear to the senses; nor is it reasonable to conclude, merely because one event, in one instance, precedes another, that therefore the one is the cause, the other the effect. . . . There is some other principle, which determines him to form such a conclusion. This principle is Custom or Habit. For wherever the repetition of any particular act or operation produces a propensity to renew the same act or operation, without being impelled by any reasoning or process of the understanding; we always say, that this propensity is the effect of Custom. . . . All inferences from experience, therefore, are effects of custom, not of reasoning. Custom, then, is the great guide of human life. It is that principle alone, which renders our experience useful to us, and makes us expect, for the future, a similar train of events with those which have appeared in the past. Without the influence of custom, we should be entirely ignorant of every matter of fact, beyond what is immediately present to the memory and senses. We should never know how to adjust means to ends, or to employ our natural powers in the production of any effect. There would be an end at once of all action, as well as of the chief part of speculation.[6]

The economist Adam Smith (1723–1790) feared the effects of repetitive work on the laboring poor.

ADAM SMITH, *AN INQUIRY INTO THE NATURE AND CAUSES OF THE WEALTH OF NATIONS* (1776)

The man whose whole life is spent in performing a few simple opera-tions, of which the effects are perhaps always the same, or very nearly the same, has no occasion to exert his understanding or to exercise his invention in finding out expedients for removing difficulties which never occur. He naturally loses, therefore, the habit of such exertion, and generally becomes as stupid and ignorant as it is possible for a human creature to become. The torpor of his mind renders him not only inca-pable of relishing or bearing a part in any rational conversation, but of conceiving any generous, noble, or tender sentiment, and consequently of forming any just judgement concerning many even of the ordinary duties of private life. Of the great and extensive interests of his country he is altogether incapable of judging, and unless very particular pains have been taken to render him otherwise, he is equally incapable of defending his country in war. The uniformity of his stationary life naturally cor-rupts the courage of his mind, and makes him regard with abhorrence the irregular, uncertain, and adventurous life of a soldier. It corrupts even the activity of his body, and renders him incapable of exerting his strength with vigour and perseverance in any other employment than that to which he has been bred. His dexterity at his own particular trade seems, in this manner, to be acquired at the expense of his intellectual, social, and martial virtues. But in every improved and civilised society this is the state into which the labouring poor, that is, the great body of the people, must necessarily fall, unless government takes some pains to prevent it.[7]

Thomas Reid (1710–1796), best known as the founder of the Scottish "Common Sense" school of philosophy, wondered whether some of our habits might also have a deeper wellspring.

THOMAS REID, "OF POWER" (1792)

I am rather inclined to think that our first exertions are instinctive, with-out any distinct conception of the event that is to follow, consequently without will to produce that event. And that finding by experience that such exertions are followed by such events, we learn to make the exertion voluntarily and deliberately, as often as we desire to produce the event. And when we know or believe that the event depends upon our exertion, we have the conception of power in ourselves to produce that event.[8]

For Immanuel Kant (1724–1804), actions could only be considered moral when motivated by duty to a categorical imperative. By this definition, habits had no ethical status.

IMMANUEL KANT, *ANTHROPOLOGY FROM A PRAGMATIC POINT OF VIEW* (1798)

Habit . . . makes the endurance of evil easy (which, under the name of patience, is falsely honored as a virtue), because sensations of the same type, when continued without alteration for a long time, draw our attention away from the senses so that we are scarcely conscious of them at all. On the other hand, habit also makes the consciousness and the remembrance of good that has been received more difficult, which then gradually leads to ingratitude (a real vice). . . . Acquired habit deprives good actions of their moral value because it undermines mental freedom and, moreover, it leads to thoughtless repetitions of the same acts (monotony), and thus becomes ridiculous.[9]

Despite favoring "reward with warmth and eloquence of approbation" in the schoolroom, the Irish writer and educationalist Maria Edgeworth (1768–1849) conceded that corporal punishment was often the only means for breaking the dominion of bad habits.

MARIA EDGEWORTH AND RICHARD LOVELL EDGEWORTH, *PRACTICAL EDUCATION* (1798)

Young people of a torpid, indolent temperament are much under the dominion of habit; if they happen to have contracted any disagreeable or bad habits, they have seldom sufficient energy to break them. The stimulus of sudden pain is necessary in this case. The pupil may be perfectly convinced that such a habit ought to be broken, and may will to break it most sincerely; but may yet be incapable of the voluntary exertion requisite to obtain success. It would be dangerous to let the habit, however insignificant, continue victorious, because the child would hence be discouraged from all future attempts to battle with himself. Either we should not attempt the conquest of the habit, or we should persist till we have vanquished. The confidence, which this sense of success will give the pupil, will probably in his own opinion be thought well worthy the price. Neither his reason nor his will was in fault; all he wanted was strength to break the diminutive chains of habit; chains which, it seems, have power to enfeeble their captives exactly in proportion to the length of time they are worn.[10]

The idea that habits were transmissible from parent to offspring, today known as Lamarckism or soft inheritance, had wide support in the eighteen and nineteenth century, but Jean-Baptiste Lamarck (1744–1829) would become its most noted champion.

JEAN-BAPTISTE LAMARCK, *ZOOLOGICAL PHILOSOPHY* (1809)

Naturalists have remarked that the structure of animals is always in perfect adaptation to their functions, and have inferred that the shape and condition of their parts have determined the use of them. Now this is a mistake: for it may be easily proved by observation that it is on the contrary the needs and uses of the parts which have caused the development of these same parts, which have even given birth to them when they did not exist, and which consequently have given rise to the condition that we find in each animal.

If this were not so, nature would have had to create as many different kinds of structure in animals, as there are different kinds of environment in which they have to live; and neither structure nor environment would ever have varied.

This is indeed far from the true order of things. If things were really so, we should not have race-horses shaped like those in England; we should not have big draught-horses so heavy and so different from the former, for none such are produced in nature; in the same way we should not have basset-hounds with crooked legs, nor grey-hounds so fleet of foot, nor water-spaniels, etc.; we should not have fowls without tails, fantail pigeons, etc.; finally, we should be able to cultivate wild plants as long as we liked in the rich and fertile soil of our gardens, without the fear of seeing them change under long cultivation.

A feeling of the truth in this respect has long existed; since the following maxim has passed into a proverb and is known by all. Habits form a second nature.[11]

French philosopher Félix Ravaisson (1813–1900) considers the inverse relationship between habit and consciousness.

FELIX RAVAISSON, *OF HABIT* (1838)

Yet what is the difference between the tendencies engendered by the continuity or repetition of action, and the primitive tendencies that constitute our nature? What is the difference between habit and instinct?

Like habit, instinct is the tendency towards an end without will and distinct consciousness. Only, instinct is more unreflective, more irresistible, more infallible. Habit draws increasingly near to, perhaps without ever attaining, the reliability, necessity and perfect spontaneity of instinct. Between habit and instinct, between habit and nature, the difference is merely one of degree, and this difference can always be lessened and reduced.

Like effort between action and passion, habit is the dividing line, or the middle term, between will and nature; but it is a moving middle term, a dividing line that is always moving, and which advances by an imperceptible progress from one extremity to the other.

Habit is thus, so to speak, the infinitesimal differential, or, the dynamic fluxion from Will to Nature. Nature is the limit of the regressive movement proper to habit.

Consequently, habit can be considered as a method—as the only real method—for the estimation, by a convergent infinite series, of the relation, real in itself but incommensurable in the understanding, of Nature and Will.

In descending gradually from the clearest regions of consciousness, habit carries with it light from those regions into the depths and dark night of nature. Habit is an acquired nature, a second nature, that has its ultimate ground in primitive nature, but which alone explains the latter to the understanding. It is, finally, a natured nature, the product and successive revelation of *naturing* nature.[12]

Though best remembered as a social Darwinist, Herbert Spencer (1820–1903) held that the "inheritance of acquired characteristics," rather than natural selection, was best placed to explain the physical and psychological peculiarities of the human race.

HERBERT SPENCER, *PRINCIPLES OF PSYCHOLOGY* (1855)

It is not simply that a modified form of constitution produced by new habits of life, is bequeathed to future generations; but it is that the modified nervous tendencies produced by such new habits of life, are also bequeathed: and if the new habits of life become permanent, the tendencies become permanent. This is illustrated in every creature respecting which we have the requisite experience, from man downwards. Though, among the families of a civilized society, the changes of occupation and habit from generation to generation, and the intermarriage of families having different occupations and habits, very greatly confuse the evidence of psychical transmission; yet, it needs but to consider national characters, in which these disturbing causes are averaged, to see distinctly, that mental peculiarities produced by habit become hereditary. We know that there are warlike, peaceful, nomadic, maritime, hunting, commercial races—races that are independent or slavish, active or slothful,—races that display great varieties of disposition; we know that many of these, if not all, have a common origin; and hence there can be no question that these varieties of disposition, which have a more or less evident relation to habits of life, have been gradually induced and established in successive generations, and have become organic. That is to say, the tendencies to certain combinations of psychical changes have become organic.[13]

The novelist Samuel Butler (1835–1902), one of the most trenchant critics of evolutionism, believed that Darwin had neglected to fully consider inherited traits as a function of biological or "unconscious" memory.

SAMUEL BUTLER, *LIFE AND HABIT* (1878)

We have seen that we cannot do anything thoroughly till we can do it unconsciously, and that we cannot do anything unconsciously till we can do it thoroughly; this at first seems illogical; but logic and consistency are luxuries for the gods, and the lower animals, only. Thus a boy cannot really know how to swim till he can swim, but he cannot swim till he knows how to swim. Conscious effort is but the process of rubbing off the rough corners from these two contradictory statements, till they eventually fit into one another so closely that it is impossible to disjoin them.

Whenever, therefore, we see any creature able to go through any complicated and difficult process with little or no effort—whether it be a bird building her nest, or a hen's egg making itself into a chicken, or an ovum turning itself into a baby—we may conclude that the creature has done the same thing on a very great number of past occasions.

We found the phenomena exhibited by heredity to be so like those of memory, and to be so utterly inexplicable on any other supposition, that it was easier to suppose them due to memory in spite of the fact that we cannot remember having recollected, than to believe that because we cannot so remember, therefore the phenomena cannot be due to memory.

We were thus led to consider "personal identity," in order to see whether there was sufficient reason for denying that the experience, which we must have clearly gained somewhere, was gained by us when we were in the persons of our forefathers; we found, not without surprise, that unless we admitted that it might be so gained, in so far as that we once actually were our remotest ancestor, we must change our ideas concerning personality altogether. We therefore assumed that the phenomena of heredity, whether as regards instinct or structure were mainly due to memory of past experiences, accumulated and fused till they had become automatic, or quasi automatic, much in the same way as after a long life—[14]

... "Old experience do attain
To something like prophetic strain."

Nietzsche considered the tyranny—and necessity—of habit.

FRIEDRICH NIETZSCHE, *THE GAY SCIENCE* (1882)

Brief habits.—I love brief habits and consider them invaluable means for getting to know many things and states down to the bottom of their sweetnesses and bitternesses; my nature is designed entirely for brief habits, even in the needs of its physical health and generally *as far as* I can see at all, from the lowest to the highest. I always believe *this* will give me lasting satisfaction—even brief habits have this faith of passion, this faith in eternity—and that I am to be envied for having found and recognized it, and now it nourishes me at noon and in the evening and spreads a deep contentment around itself and into me, so that I desire nothing else, without having to compare, despise, or hate. And one day its time is up; the good thing parts from me, not as something that now disgusts me but peacefully and sated with me, as I with it, and as if we ought to be grateful to each other and so shake hands to say farewell. And already the new waits at the door along with my faith— the indestructible fool and sage!—that this new thing will be the right thing, the last right thing. This happens to me with dishes, thoughts, people, cities, poems, music, doctrines, daily schedules, and ways of living. *Enduring* habits, however, I hate, and feel as if a tyrant has come near me and the air around me is *thickening* when events take a shape that seems inevitably to produce enduring habits—for instance, owing to an official position, constant relations with the same people, a permanent residence, or uniquely good health. Yes, at the very bottom of my soul I am grateful to all my misery and illnesses and whatever is imperfect in me because they provide a hundred back doors through which I can escape enduring habits. To me the most intolerable, the truly terrible, would of course be a life entirely without habits, a life that continually demanded improvisation—that would be my exile and my Siberia.[15]

William James advised can-do America to make friends with the "effortless custody of automatism."

WILLIAM JAMES, *PSYCHOLOGY: BRIEFER COURSE* (1892)

If the period between twenty and thirty is the critical one in the formation of intellectual and professional habits, the period below twenty is more important still for the fixing of personal habits, properly so called, such as vocalization and pronunciation, gesture, motion, and address. Hardly ever is a language learned after twenty spoken without a foreign accent; hardly ever can a youth transferred to the society of his betters unlearn the nasality and other vices of speech bred in him by the associations of his growing years. Hardly ever, indeed, no matter how much money there be in his pocket, can he even learn to *dress* like a gentleman-born. The merchants offer their wares as eagerly to him as to the veriest 'swell' but he simply *cannot* buy the right things. An invisible law, as strong as gravitation, keeps him within his orbit, arrayed this year as he was the last; and how his better-clad acquaintances contrive to get the things they wear will before him a mystery till his dying day.

The great thing, then, in all education, is to *make our nervous system our ally instead of our enemy*. It is to fund and capitalize our acquisitions, and live at ease upon the interest of the fund. *For this we must make automatic and habitual, as early as possible, as many useful actions as we can*, and guard against the growing into ways that are likely to be disadvantageous to us, as we should guard against the plague. The more of the details of our daily life we can hand over to the effortless custody of automatism, the more our higher powers of mind will be set free for their own proper work. There is no more miserable human being than one in whom nothing is habitual but indecision, and for whom the lighting of every cigar, the drinking of every cup, the time of rising and going to bed every day, and the beginning of every bit of work, are subjects of express volitional deliberation. Full half the time of such a man goes to the deciding, or regretting, of matters which ought to be so ingrained in him as practically not to exist for his consciousness at all. If there be such daily duties not yet ingrained in any one of my readers, let him begin this very hour to set the matter right.[16]

Psychologist Edward L. Thorndike (1874–1949) suggested that repetition alone is not a sufficient foundation for building good habits.

EDWARD L. THORNDIKE, *THE PRINCIPLES OF TEACHING BASED ON PSYCHOLOGY* (1906)

It is a fundamental law of mental life that if a mental state or bodily act is made to follow or accompany a certain situation with resulting satisfaction it will tend to go with that situation in the future. The applications of the law to teaching are comprised in the simple and obvious, but too commonly neglected rules. Put together what you wish to have go together. Reward good impulses. Conversely; *Keep apart what you wish to have separate. Let undesirable impulses bring discomfort.*

Obvious as these rules are to the student of human nature, they are constantly violated. For instance, the commonest school punishment of the writer's schooldays was to keep pupils in school over-time, thus putting the idea of punishment, of undesirability, into closest connection with the experience of school and school work. . . .

Still more flagrant are the violations of the law, *Reward good impulses.* The mother neglects her children when they are quiet and decent and plays with them only when they cry. Consequently there are many crying babies. The child is refused a favor when he asks once, but if he teases a score of times it is finally granted. Consequently there are many teasing boys.

The mind does not gravitate toward truth, wisdom and goodness of its own accord. What it does is to keep together those ideas, feelings and acts which it experiences together and with resulting satisfaction. If a situation is to call up its proper response, that response should be put with that situation in the individual's experience. The mind does not do something for nothing. If it is to drop one tendency and cherish another, the latter must be made the more satisfying. To get habits we must make them and reward them. *Put together what you wish to have go together. Reward good impulses.*[17]

Ivan Pavlov's (1849–1936) research on conditioned reflexes left a huge impact beyond medicine and physiology, with American behaviorism adopting the concept as a master key for understanding the psychology of learning and habit formation.

IVAN P. PAVLOV, *CONDITIONED REFLEXES* (1927)

Food, through its chemical and physical properties, evokes the salivary reflex in every dog right from birth, whereas this new type claimed as reflex—"the signal reflex"—is built up gradually in the course of the animal's own individual existence. But can this be considered as a fundamental point of difference, and can it hold as a valid argument against employing the term "reflex" for this new group of phenomena? It is certainly a sufficient argument for making a definite distinction between the two types of reflex and for considering the signal reflex in a group distinct from the inborn reflex. But this does not invalidate in any way our right logically to term both "reflex," since the point of distinction does not concern the character of the response on the part of the organism, but only the mode of formation of the reflex mechanism. We may take the telephonic installation as an illustration. Communication can be effected in two ways. My residence may be connected directly with the laboratory by a private line, and I may call up the laboratory whenever it pleases me to do so; or on the other hand, a connection may have to be made through the central exchange. But the result in both cases is the same. The only point of distinction between the methods is that the private line provides a permanent and readily available cable, while the other line necessitates a preliminary central connection being established. In the one case the communicating wire is always complete, in the other case a small addition must be made to the wire at the central exchange. We have a similar state of affairs in reflex action. The path of the inborn reflex is already completed at birth; but the path of the signalizing reflex has still to be completed in the higher nervous centres. We are thus brought to consider the mode of formation of new reflex mechanisms. A new reflex is formed inevitably under a given set of physiological conditions, and with the greatest ease, so that there is no need to take the subjective states of the dog into consideration. With a complete understanding of all the factors involved, the new signalizing reflexes are under the absolute control of the experimenter; they proceed according to as rigid laws as do any other physiological processes, and must be regarded as being in every sense a part of the physiological activity of living beings. I have termed this new group of reflexes **conditioned reflexes** to distinguish them from the inborn or **unconditioned reflexes**.[18]

John Broadus Watson's (1878–1958) "behaviorist manifesto" tied the dream of psychology as natural science to the fantasy of it furnishing "the educator, the physician, the jurist and the business man" with a means of understanding and controlling human action.

JOHN B. WATSON, *PSYCHOLOGY AS THE BEHAVIORIST VIEWS IT* (1913)

The psychology which I should attempt to build up would take as a starting point, first, the observable fact that organisms, man and animal alike, do adjust themselves to their environment by means of hereditary and habit equipments. These adjustments may be very adequate or they may be so inadequate that the organism barely maintains its existence; secondly, that certain stimuli lead the organisms to make the responses. In a system of psychology completely worked out, given the response the stimuli can be predicted; given the stimuli the response can be predicted. Such a set of statements is crass and raw in the extreme, as all such generalizations must be. Yet they are hardly more raw and less realizable than the ones which appear in the psychology texts of the day. I possibly might illustrate my point better by choosing an everyday problem which anyone is likely to meet in the course of his work. Some time ago I was called upon to make a study of certain species of birds. Until I went to Tortugas I had never seen these birds alive. When I reached there I found the animals doing certain things: some of the acts seemed to work peculiarly well in such an environment, while others seemed to be unsuited to their type of life. I first studied the responses of the group as a whole and later those of individuals. In order to understand more thoroughly the relation between what was habit and what was hereditary in these responses, I took the young birds and reared them. In this way I was able to study the order of appearance of hereditary adjustments and their complexity, and later the beginnings of habit formation. My efforts in determining the stimuli which called forth such adjustments were crude indeed. Consequently my attempts to control behavior and to produce responses at will did not meet with much success. Their food and water, sex and other social relations, light and temperature conditions were all beyond control in a field study. I did find it possible to control their reactions in a measure by using the nest and egg (or young) as stimuli. It is not necessary in this paper to develop further how such a study should be carried out and how work of this kind must be supplemented by carefully controlled laboratory experiments. Had I been called upon to examine the natives of some of the Australian tribes, I should have gone about my task in the same way. I should have found the problem more difficult: the types of responses called forth by physical stimuli would have been more varied, and the number of effective stimuli larger. I should have had to determine the social setting of their lives in a far more careful way. These savages would be more influenced by the responses of each other than was the case with the birds. Furthermore, habits would have been more complex and the influences of past habits upon the present responses would have appeared more clearly. Finally, if I had been called upon to work out the psychology of the educated European, my problem would have required several lifetimes. But in the one I have at my disposal I should have followed the same general line of attack. In the main, my desire in all such work is to gain an accurate knowledge of adjustments and the stimuli calling them forth. My final reason for this is to learn general and particular methods by which I may control behavior. . . .

It has been shown that improvement in habit comes unconsciously. The first we know of it is when it is achieved—when it becomes an object. I believe that 'consciousness' has just as little to do with improvement in thought processes. Since, according to my view, thought processes are really motor habits in the larynx, improvements, short cuts, changes, etc., in these habits are brought about in the same way that such changes are produced in other motor habits.[19]

Emil Kraepelin (1856–1926), the father of modern-day psychiatry, considers the nature of repetitive motor actions in patients suffering from Dementia Praecox (schizophrenia).

EMIL KRAEPELIN, *DEMENTIA PRAECOX AND PARAPHRENIA* (1909)

Sometimes the whole volitional expression of the patient is dominated by stereotypies for a long time, so that his doings resolve themselves into an almost uninterrupted series of senseless movements which are either monotonous, or repeat themselves with slight changes. A certain rhythm invariably results. The patients rock themselves from one leg on to the other, keep time, "pull letters away from their fingertips," spread out their fingers with a quavering movement, clap their hands, shake their heads, bellow keeping time, give themselves boxes on their ears, run up and down in double quick time. About the motives for these proceedings, no satisfactory account is got from them. A patient who always rocked himself rhythmically from side to side, simply explained, "It happens so in me," "I must shake my head or else I am in terror," "I must constantly say things," "I must scream without wanting to, there is that impulse in me," "I must throw myself about at night in bed as if a strange power threw me," "I must turn round, as when a magnet draws a needle," "I could not have rested till I had done that," are similar expressions.

We may well suppose that also the development of such stereotypies, which later give such a peculiar appearance to the terminal states of the disease and likewise to many forms of idiocy, is specially favoured by the failure of healthy volitional impulses, perhaps first made possible. Many experiences at least indicate that the mechanism of our will possesses arrangements acquired long ago, which favour a rhythmical repetition of the same discharges; their influence will be able to make itself felt as soon as the impulses disappear which serve for a realisation of intentions.[20]

Psychologist and philosopher John Dewey (1859–1952), one of the most astute critics of the stimulus-response theory of behavior, thought that habits were more dynamic and extensive than either Watson or his followers allowed.

JOHN DEWEY, *HUMAN NATURE AND CONDUCT* (1922)

Every habit creates an unconscious expectation. It forms a certain outlook. What psychologists have laboriously treated under the caption of association of ideas has little to do with ideas and everything to do with the influence of habit upon recollection and perception. A habit, a routine habit, when interfered with generates uneasiness, sets up a protest in favor of restoration and a sense of need of some expiatory act or else it goes off in casual reminiscence. It is the essence of routine to insist upon its own continuation. Breach of it is violation of right. Deviation from it is transgression.

All that metaphysics has said about the *nisus* of Being to conserve its essence and all that a mythological psychology has said about a special instinct of self-preservation is a cover for the persistent self-assertion of habit. Habit is energy organized in certain channels. When interfered with, it swells as resentment and as an avenging force. To say that it will be obeyed, that custom makes law, that nomos is lord of all, is after all only to say that habit is habit.[21]

In Svevo's (1861–1928) classic novel Zeno's Conscience, *Zeno Cosino, a neurotic businessman in bondage to his "tobacco habit," wonders whether a lifetime of "last" cigarettes might have evolved into a psychological cover story.*

ITALO SVEVO, *ZENO'S CONSCIENCE* (1923)

Now that I am here, analyzing myself, I am seized by a suspicion: Did I perhaps love cigarettes so much because they enabled me to blame them for my clumsiness? Who knows? If I had stopped smoking, would I have become the strong, ideal man I expected to be? Perhaps it was this suspicion that bound me to my habit, for it is comfortable to live in the belief that you are great, though your greatness is latent. I venture this hypothesis to explain my youthful weakness, but without any firm conviction. Now that I am old and no one demands anything of me, I still pass from cigarette to resolve, and from resolve to cigarette. What do those resolutions mean today?[22]

Bad habits became big business. In the expanding marketplace for techniques, courses and treatments aimed at eradicating fixed habits, F. M. Alexander's (1869–1955) kinesthetic approach, based on control of motor habits, gained many notable followers, including John Dewey, Aldous Huxley, and George Bernard Shaw.

F. M. ALEXANDER, *CONSTRUCTIVE CONSCIOUS CONTROL OF THE INDIVIDUAL* (1923)

In this matter of bad habits, and the lack of control which they connote, we must recognize the fact that the human creature cannot be expected to exercise control in the different spheres of his activity in civilization unless he is in possession of reliable sensory appreciation and of a satisfactory use of the psycho-physical mechanisms involved. People who are lacking in control will be found to be imperfectly co-ordinated, and their sensory appreciation to be unreliable, and no form of discipline or other outside influence can secure that satisfactory standard of psycho-physical functioning without which the individual cannot command a satisfactory standard of control within or without the organism.

Where the human being manifests this lack of control, he needs to be re-educated on a general basis so that reliable sensory appreciation may be restored, together with a satisfactory employment of the psycho-physical mechanisms. The processes of this form of re-education demand that the "means-whereby" to any "end" must be reasoned out, not on a specific but on a general basis, and with the continued use of these processes of reasoning, uncontrolled impulses and "emotional gusts" will gradually cease to dominate, and will ultimately be dominated. The organism will not then be called upon to satisfy those unhealthy cravings which we find associated with unreliable and delusive sensory appreciation (debauched kinesthesia). . . .

Worry is one of these bad habits which, once established, are very hard to break. A curious feature of this habit is that, in certain cases, though you may remove the cause for worry, and the subject may admit that the cause has been removed, the removal of the cause does not remove the "mental" state which the subject declared was the cause of the worry. The fact is, the person has developed the worry habit, a state in which he manufactures the stimulus to worry. . . .

[I]n any consideration of "mental" and "physical" phenomena it must be remembered that in our present stage of evolution on the subconscious plane, the response to any stimulus or stimuli is at least seventy-five per cent subconscious response (chiefly feeling) as against twenty-five per cent any other response, this estimate of the ratio of subconscious response being probably too low. When these facts are fully realized by all those who are interested in education and in the conduct of life generally, there may be some chance of the realization of those commendable ideals for the uplifting of mankind cherished by leaders in the social, religious, and political spheres.[23]

The Swiss child psychologist Jean Piaget (1896–1980) asked whether our early sensorimotor habits might pave the way for intelligent behavior.

JEAN PIAGET, *THE ORIGINS OF INTELLIGENCE IN CHILDREN* (1936)

Sucking [of the] thumb or tongue, following with the eyes moving objects, searching for where sounds come from, grasping solid objects to suck or look at them, etc., are the first habits which appear in the human being. We have described their appearance in detail but the question may be asked in a general way, what sensorimotor habit is and how it is constituted. Furthermore, and it is with this sole aim that we have studied the first acquired adaptations, it may be asked in what way habitual association prepares the intelligence and what the relationships are between these two types of behavior patterns. Let us begin with this last point. In psychology there has always been a tendency to trace back the active operations of intelligence to the passive mechanisms arising from association or habit. To reduce the causal link to a matter of habit, to reduce the generalization characteristic of the concept to the progressive application of habitual systems, to reduce judgment to an association, etc. such are the common positions of a certain psychology dating from Hume and Bain. The idea of the conditioned reflex, which is perhaps misused today, undoubtedly revives the terms of the problem, but its application to psychology certainly remains in the prolongation of this tradition. Habit, too, has always seemed to some people to be the antithesis of intelligence. Where the second is active invention, the first remains passive repetition; where the second is awareness of the problem and an attempt at comprehension, the first remains tainted with lack of awareness and inertia, etc. The solution we shall give to the question of intelligence . . . is to consider the formation of habits as being due to an activity whose analogies with intelligence are purely functional. . . . Association and habit form the automatization of an activity which functionally prepares intelligence.[24]

For the philosopher Maurice Merleau-Ponty (1908–1961), habit was above all else a form of embodiment.

MAURICE MERLEAU-PONTY, *THE PHENOMENOLOGY OF PERCEPTION* (1945)

To habituate oneself to a hat, an automobile, or a cane is to take up residence in them, or inversely, to make them participate within the voluminosity of one's own body. Habit expresses the power we have of dilating our being in the world, or of altering our existence through incorporating new instruments. One can know how to type without knowing how to indicate where on the keyboard the letters that compose the words are located. Knowing how to type, then, is not the same as knowing the location of each letter on the keyboard, nor even having acquired a conditioned reflex for each letter that is triggered upon seeing it.

But if habit is neither a form of knowledge nor an automatic reflex, then what is it? It is a question of a knowledge in our hands, which is only given through a bodily effort and cannot be translated by an objective designation. The subject knows where the letters are on the keyboard just as we know where one of our limbs is—a knowledge of familiarity that does not provide us with a position in objective space. The movement of his fingers is not presented to the typist as a spatial trajectory that can be described, but merely as a certain modulation of motricity, distinguished from every other through its physiognomy. . . .

But the phenomenon of habit in fact leads us to rework our notion of "understanding" and our notion of the body. To understand is to experience [éprouver] the accord between what we aim at and what is given, between the intention and the realization—and the body is our anchorage in a world.[25]

In The Concept of Mind, *his seminal attack on Cartesian dualism, the philosopher Gilbert Ryle (1900–1976) questioned the status of habitual acts, asking to what extent they fell outside the ambit of intelligent or purposive behavior.*

GILBERT RYLE, *CONCEPT OF MIND* (1949)

When we say that someone acts in a certain way from sheer force of habit, part of what we have in mind is this, that in similar circumstances he always acts in just this way; that he acts in this way whether or not he is attending to what he is doing; that he is not exercising care or trying to correct or improve his performance; and that he may, after the act is over, be quite unaware that he has done it. Such actions are often given the metaphorical title 'automatic'. Automatic habits are often deliberately inculcated by sheer drill, and only by some counter-drill is a formed habit eradicated.

But when we say that someone acts in a certain way from ambition or sense of justice, we mean by implication to deny that the action was merely automatic. In particular we imply that the agent was in some way thinking or heeding what he was doing, and would not have acted in that way, if he had not been thinking what he was doing. But the precise force of this expression 'thinking what he was doing' is somewhat elusive. I certainly can run upstairs two stairs at a time from force of habit and at the same time notice that I am doing so and even consider how the act is done. I can be a spectator of my habitual and of my reflex actions and even a diagnostician of them, without these actions ceasing to be automatic. Notoriously such attention sometimes upsets the automatism.[26]

Philosopher of science Raymond Ruyer (1902–1987), best known for his work Neofinalism, *drew on findings from embryology to propose that habits could never assume the force or tenor that the neo-Lamarckians continued to ascribe them.*

RAYMOND RUYER, *NEOFINALISM* (1952)

Can the habit of sucking create the instinct of sucking, the instinct of swallowing, of digesting and of forming a stomach and a digestive tract for oneself? Can the habit of making provisions create the instinct of hoarding, then the formative instinct of organic reserves of sugar or fat? Can the sexual habits of a male individual create the instincts, then the sexual organs, of the male? And "whose" habit is it that harmonizes the instincts and the organs of the male and female? It is indeed clear that psycho-Lamarckism inverts the real order. If our tools are akin to external organs, and vice versa, if our organs are akin to tools (for the problems of nature, the order of comparison matters little), it is in fact the tools that presuppose the existence of organs and not the reverse (for the problems of origin, the order by contrast matters a great deal).

Neo-Lamarckians were deceived by the phenomenon of passage from the conscious to the unconscious, which seems to bring habit close to instinct. But active habits, at least within the bounds of our experience, become "unconscious" only in the manner of a psychological "other-I." They remain in the domain of the psychological in the ordinary sense of the term. A habit never assumes the character of an instinct nor especially the character of an instinct that forms organs; it never enters into the region of biological autosubjectivity. Secondary consciousness is never transformed into primary consciousness. . . .

Habit, learning, cannot be the primary element of organic finality nor, moreover, of finality in general. Habit is an auxiliary of finality, an accessory channeling, an accommodation of subordinated details. Isolated from a principle of higher finality, habit always risks losing the general point of view to confine itself to a small domain of accommodation, often by creating adverse step-by-step "adhesions."[27]

Anthropologist Claude Levi-Strauss (1908–2009) observed that habit and customs were universally justified by "secondary elaborations" that neglected their psychological import.

CLAUDE LÉVI-STRAUSS, *STRUCTURAL ANTHROPOLOGY* (1958)

We know that among most primitive peoples it is very difficult to obtain a moral justification or a rational explanation for any custom or institution. When he is questioned, the native merely answers that things have always been this way, that such was the command of the gods or the teaching of the ancestors. Even when interpretations are offered, they always have the character of rationalizations or secondary elaborations. There is rarely any doubt that the unconscious reasons for practicing a custom or sharing a belief are remote from the reasons given to justify them. Even in our own society, table manners, social etiquette, fashions of dress, and many of our moral, political, and religious attitudes are scrupulously observed by everyone, although their real origin and function are not often critically examined. We act and think according to habit, and the extraordinary resistance offered to even minimal departures from custom is due more to inertia than to any conscious desire to maintain usages which have a clear function.[28]

The laboratory studies of Harvard social psychologist Ellen Langer (1947–) found that "mindless" or habit-driven behavior was the default setting for much of our daily interactions.

ELLEN LANGER, ARTHUR BLANK, AND BENZION CHANOWITZ, "THE MINDLESSNESS OF OSTENSIBLY THOUGHTFUL ACTION" (1978)

Social psychology is replete with theories that take for granted the "fact" that people think. . . . [Our own] studies taken together support the contention that when the structure of a communication, be it oral or written, semantically sound or senseless, is congruent with one's past experience, it may occasion behavior mindless of relevant details. Clearly, some information from the situation must be processed in order for a script to be cued. However, what is being suggested here is that only a minimal amount of structural information may be attended to and that this information may not be the most useful part of the information available. . . .

When does this mindless activity take place? If the interpretation offered for these studies is correct, then it would suggest that the occurrence may not be infrequent nor restricted to overlearned motoric behavior like typewriting. Instead, if complex verbal interactions can be overlearned, mindlessness may indeed be the most common mode of social interaction. While such mindlessness may at times be troublesome, this degree of selective attention, of tuning the external world out, may be an achievement.[29]

While trying to understand the behavior of stroke patients whose actions seemed dictated almost entirely by environmental cues and stimuli, François Lhermitte (1921–1998), a neurologist at Salpêtrière Hospital in Paris, began to understand the nature of unconscious priming in our daily lives.

JOHN BARGH, *BEFORE YOU KNOW IT* (2017)

Lhermitte started simply. Filling two glasses with water, he set them down in front of the patients, who promptly drank them right down. Nothing unusual there, of course. Except Lhermitte kept filling the glasses, and the patients kept drinking them all right down, glass after glass, even while complaining about being painfully full. They could not help but drink the full glasses of water placed in front of them. On a different occasion, the doctor took the man to his home, an apartment. He led the man out onto his balcony, which overlooked a nearby park, and they admired the view together. Right before reentering the apartment, Lhermitte softly said "museum," and when back inside the patient proceeded to scrutinize the paintings and posters on the walls with great interest, also lavishing his attention on common objects that sat on the tables—plates and cups with little aesthetic interest—as if they too were actual works of art. On next being shown the bedroom, the man looked at the bed, proceeded to undress, and got into it. Soon he was asleep.

What was going on here? They appeared unconsciously compelled by the naturally occurring primes in their environment, yet they had no trouble consciously justifying all these activities—their water-chugging, art appreciation, gardening. . . . The brain's fine-tuned responses learned in the past—or guided by the future, in relation to any plans or goals they might have had—had been replaced by a hypersensitivity to the present, and seemingly only the present. . . .

Every human mind, then, is a kind of mirror, generating potential behaviors that reflect back the situations and environments in which we find ourselves—a glass of water says "drink me," flower beds say "tend me," beds say "sleep in me," museums say "admire me."[30]

But what happens to a habit when its cues disappear?

WENDY WOOD, *GOOD HABITS, BAD HABITS* (2019)

For two days in the dreary late London winter of 2014, the Underground metro system shut down. . . . The disruption was monumental. On the days of the strike, only about 60 percent of these commuters were able to enter at their normal stations, and about 50 percent exited as usual. In between, commuters were improvising. Surprisingly, the collective improvisations didn't drastically increase commuting time. On average, people spent only 6 percent more time in transit. Some people actually got to work faster—especially the commuters who typically used slow lines or traveled on distorted areas of the map.

Of course, commuters could have experimented with alternative routes even without a strike. Only their habits were stopping them from trying different Tube lines or starting or stopping at different stations. But in the rush of daily life, we don't often take the time to experiment. We find something that works adequately, and we stick with it. For the sake of ease, we settle.

The closure of the Underground made this "adequate" way of doing things briefly impossible. This is called *habit discontinuity*—a term coined by researcher Bas Verplanken to describe how our habits are disrupted by changes in context. When habitual cues disappear, we can no longer respond automatically. We have to make conscious decisions. We are open to change—even, sometimes, serendipitously finding improvement.[31]

7

THE DIVIDED SELF: CLINICAL AND EXPERIMENTAL INSIGHTS

"We are all subjected to two distinct natures in the same person," observes Robert Wigham, the despairing anti-hero in James Hogg's 1824 novel *The Private Memoirs and Confessions of a Justified Sinner*. "I myself have suffered grievously in that way. The spirit that now directs my energies is not that with which I was endowed at my creation. It is changed within me, and so is my whole nature."[1]

While the divided selves and spectral doubles that feature in so many nineteenth-century novels and short stories were essentially versions of the double-minded man invoked in James' epistles, Hogg's *Confessions* spoke to a new and more profound form of psychological dislocation: double consciousness. Most often connected to somnambulism, hysteria and epilepsy, double consciousness was a pre-eminently feminine condition, characterized by the emergence of a secondary personality, with amnesia. The century's best-known clinical cases of double consciousness, from Rachel Baker to Felida X, were clearly taking flight from reality, absenting themselves from unfulfilled lives, but those afflicted by double consciousness were also experimental subjects, providing alienists and psychiatrists, often using mesmerism or hypnosis, with evidence of the unconscious in crisis. In this respect, double consciousness was largely *iatrogenic*: a condition caused or stimulated by medical examination or suggestion.

It was not long after the reporting of the first cases of double conscious in America that the British physician Arthur Ladbroke Wigan proposed, in his book *A New View of Insanity*, that a duality of the self was inherent in the structure of the brain. Though Wigan's book very briefly referred to some cases of damage to the corpus callosum (the largest of the commissures which provide a neuronal bridge between the two brain hemispheres), the first significant evidence of hemispheric specialization came from much later studies of disorders of movement and language by German neurologists such as Hugo Liepmann.

Fin-de-siècle studies of aphasia and apraxia strongly suggested that disconnection of the corpus callosum underpinned motor and language disorders, but the idea that the integrity of the self depended on a large white tract that allowed the two sides of the brain to remain in concert was still met with some resistance. The British psychologist William McDougall, never one to buy into a growing orthodoxy, went as far as to urge the physiologist Charles Sherrington to operate and cut through his corpus callosum should he ever be faced with an incurable illness. "If the physiologists are right the result should be a split personality. If I am right my consciousness will remain a unitary consciousness."[2]

Half a century on, William van Wagenen, a neurosurgeon at the University of Rochester Medical Center in Rochester, New York, elected to section the corpus callosum of a number of patients with intractable epilepsy in the hope of reducing their seizures. The landmark operation was to some degree successful, creating a firebreak against further seizures, but the dire psychological effects observed by Liepmann and other clinicians were not reported.

Two decades later, Roger Sperry and his colleagues at the California Institute of Technology offered a more arresting picture of the divided brain, his research on functional specialization of the cerebral hemispheres eventually earning him a Nobel Prize. One of the first split-brain patients studied in Sperry's lab developed alien hand syndrome, his left hand routinely trying to interrupt, thwart, or undo what the right was trying to do. Based on further experiments on language recognition and speech articulation, it became obvious to Sperry that split-brain patients did indeed carry "two separate spheres of conscious awareness . . . each with its own sensations, perceptions, cognitive processes, learning experiences, memories, and so on."[3]

These and other findings on split-brain research gave life to a rich seam of pseudoscience, a New Age pedagogy that tended to reduce the complexities of hemispheric laterization to a list of specious watchwords. Left brain: active, conscious, inductive, logical. Right brain: passive, subconscious, musical, and artistic. What was needed, according to the wide-eyed sponsors of this mythology, was a new marriage of left brain and right brain, a union of intellect and feeling, of verbal and intuitive thinking. With the correct education, every two of us could become one.

America's West Coast was by now the epicenter of another field of research on the splitting of the self: the multiple personality movement. Having cataloged around 50 cases of multiple personality with attendant "alters," the Santa Cruz psychiatrist Ralph Allison successfully lobbied to bring "multiple personality disorder" into the 1980 edition of American Psychiatric Association's *Diagnostic and Statistical Manual*. What followed next was a full-blown epidemic of dissociation (20,000 cases diagnosed in the United States before 1990) in which the status of the unconscious and the methods employed by its therapeutic investigators were placed on trial. Was dissociation a natural response to trauma and abuse? Or was this a psychotherapeutic drama in which patients were merely encouraged to see themselves as possessing multiple selves?

*

The terms double consciousness, divided consciousness, and double personality gained medical currency in the early decades of the nineteenth century, often applied to cases of natural somnambulism which occurred around puberty and typically involved paroxysms that gave life to an upskilled secondary personality.

HENRY DEWAR, "REPORT ON A COMMUNICATION FROM DR. DYCE OF ABERDEEN" (1823)

The communication received from Dr DYCE chiefly consists of a description of a singular affection of the nervous system, and mental powers, to which a girl of sixteen was subject immediately before puberty, and which disappeared when that state was fully established. It exemplifies the powerful influence of the state of the uterus on the mental faculties; but its chief value arises from some curious relations which it presents to the phenomena of mind, and which claim the attention of the practical metaphysician. The mental symptoms of this affection are among the number of those which are considered as uncommonly difficult of explanation. It is a case of mental disease, attended with some advantageous manifestations of the intellectual powers; and these manifestations disappearing in the same individual in the healthy state. It is an instance of a phenomenon which is sometimes called double consciousness, but is more properly a divided consciousness, or double personality; exhibiting in some measure two separate and independent trains of thought, and two independent mental capabilities, in the same individual; each train of thought, and each capability, being wholly dissevered from the other, and the two states in which they respectively predominate subject to frequent interchanges and alternations. . . .

The strong contrast between the mental states of this person under her fit, and when it was off, is to be classed with a set of facts, of which some other examples have lately come to the public knowledge. One of them was in an apparently simple girl in the neighbourhood of Stirling, who, in her sleep, talked like a profound philosopher, solved geographical problems, and enlarged on the principles of astronomy, detailing the workings of ideas which had been suggested to her mind, by over-hearing the lessons which were given by a tutor to the children of the family in which she lived. The originality of the language which she used, shewed something more than a bare repetition of what she had heard. She explained the alternations of winter and summer, for instance, by saying, that "the earth was set a-gee." Another case was mentioned in some of the newspapers, two or perhaps three years ago, of a more marked instance of double consciousness. The individual was liable to two states, each of which, if I rightly recollect continued for two or more years. In the one state, when it first came on, there was an oblivion of all former education, but no deficiency of mental vigour as applied to ideas or pursuits subsequently presenting themselves. It was necessary for this woman to recommence the studies of reading and the art of writing. A separate set of notions, and separate accomplishments were now formed. In one of the states an exquisite talent for music, and some others which implied refinement, were displayed. When another mental revolution arrived, these utterly disappeared and the individual was reduced to a level with the rest of mankind, displaying a sufficient portion of common sense; but nothing brilliant.[4]

Mary Reynolds, the first American case of double consciousness, began to commute between two states of being at the age of eighteen, in 1811. Here, she describes the onset of her duality in her own words.

SILAS WEIR MITCHELL, *MARY REYNOLDS: A CASE OF DOUBLE CONSCIOUSNESS* (1889)

These transitions always took place in my sleep. In passing from my second to my first state nothing was particularly noticeable in my sleep, but in passing from first to second state my sleep was so profound that no one could awake me, and it not unfrequently continued eighteen or twenty hours. I had generally some presentiment of the change for several days before the event. My sufferings in the near prospect of the transition from either the one or the other state were extreme, particularly from the first to the second state. When about to undergo the change, fearing I should never revert so as to know again in this world those who were dear to me, my feelings in this respect were not unlike one who was about to be separated by death, though in the second state I did not anticipate the change with such distressing apprehension as in the first. I was naturally cheerful, but more so at that time than in my natural state. I felt perfectly free from any trouble when in my second state; and for some time after I had been in that situation my feelings were such that had all my friends been laying dead beside me I do not think that it would have caused me one moment's pain of mind. At that time my feelings were never moved, either with the manifestation of joy or sorrow. I had no idea either of the past or future; nothing but the present occupied my mind.[5]

Echoing the speculations of many phrenologists before him, Brighton physician Arthur Ladbroke Wigan (1785–1847) maintained that the brain's hemispheres housed separate thoughts and independent wills.

ARTHUR LADBROKE WIGAN, *THE DUALITY OF THE MIND* (1844)

In entering on the subject of the duality of the mind and its organs, I must begin by demanding the temporary assent to certain propositions of which I am hereafter to furnish the proofs.

I believe then able to prove—

1. That each cerebrum is a distinct and perfect whole, as an organ of thought.
2. That a separate and distinct process of thinking or ratiocination may be carried on in each cerebrum simultaneously.
3. That each cerebrum is capable of a distinct and separate volition, and that these are very often opposing volitions.
4. That, in the healthy brain, one of the cerebra is almost always superior in power to the other, and capable of exercising control over the volitions of its fellow, and of preventing them from passing into acts, or from being manifested to others.
5. That when one of these cerebra becomes the subject of functional disorder, or of positive change or structure, of such a kind as to vitiate mind or induce insanity, the healthy organ can still, up to a certain point, control the morbid volitions of its fellow.
6. That this point depends partly on the extent of the disease or disorder, and partly on the degree of cultivation of the general brain in the art of self-government.
7. That when the disease or disorder of one cerebrum becomes sufficiently aggravated to defy the control of the other, the case is then one of the commonest forms of mental derangement or insanity; and that a lesser degree of discrepancy between the functions of the two cerebra constitutes the state of conscious delusion.
8. That in the insane, it is almost always possible to trace the intermixture of two synchronous trains of thought, and that it is the irregularly alternate utterance of portions of these two trains of thought which constitutes incoherence.
9. That of the two many distinct simultaneous trains of thought, one may be rational and the other irrational, or both may be irrational; but that, in either case, the effect is the same, to deprive the discourse of coherence or congruity.

Even in furious mania, this double process may be generally perceived; often it takes the form of a colloquy between the diseased mind and the healthy one, and sometimes even resembles the steady continuous argument or narrative of a sane man, more or less frequently interrupted by a madman; but persevering with tenacity of purpose in the endeavour to overpower the intruder.[6]

Educator and clergyman Francis Wayland (1796–1865) reflects on the merging disconnected states of consciousness.

FRANCIS WAYLAND, *ELEMENTS OF INTELLECTUAL PHILOSOPHY* (1854)

There have been observed occasionally abnormal cases of what may be termed double consciousness. In such a case, the present existence of the individual is at one time connected with one period of his life, and at another time with another. A young woman in Springfield, Mass, some years since, was affected in this manner. . . . Whatever she learned in the abnormal state was entirely forgotten as soon as she passed from this state to the other, but was perfectly remembered as soon as the abnormal state returned. Thus she was taught to play backgammon in both states. What she learned in the abnormal state was entirely disconnected from what she learned in her natural state, and vice versa. The acquisition made in one state was lost as soon as she entered the other and it was remarked that she learned more rapidly in the abnormal than in the normal state. The first symptom of her recovery was the blending together of the knowledge acquired in these separate conditions. As the cure advanced, they became more and more identified until the testimony of consciousness became uninterrupted and then the abnormal state vanished altogether.[7]

Fräulein Anna O. (real name Bertha Pappenheim) began treatment for hysteria with Josef Breuer (1842–1925) in 1880.

SIGMUND FREUD AND JOSEF BREUER, *STUDIES IN HYSTERIA* (1893)

When the patient (Anna O.) had been confined to her bed, and her consciousness was continually oscillating between the normal and the 'second' state, the army of hysterical symptoms that had arisen individually, and until then been latent, manifested themselves as chronic symptoms.

These were then joined by yet another group of phenomena that seemed to be of a different origin, namely the paralytic contracture of the left extremities and the paresis of the muscles supporting her head. . . .

During the entire course of the illness the two states of consciousness existed alongside one another: the primary state, in which the patient was psychically quite normal, and the 'second' state, which might well be compared with dreams, given its wealth of phantasms and hallucinations, the large gaps in her memory, the lack of inhibition and of control over her thoughts. . . . It is my opinion that the sharp division between the two states in the case of our patient simply clarifies behaviour, which is the cause of much puzzlement in many other hysterics too. In the case of Anna O. it was particularly striking to see the degree to which the products of the 'bad self', as she herself called it, influenced her moral disposition. Had they not constantly been cleared away she would have become a hysteric of the malicious variety, unruly, lethargic, ill-natured and spiteful. . . . But however sharply these two states were separated, the 'second state' did not simply jut over into the first; rather it was frequently the case that, even in very bad states, a keen and quiet observer was, as the patient put it, sitting in some corner of her mind, watching all the mad things going on. . . .

One definitely had the impression that the whole host of products from the second state, which had been slumbering, were now forcing their way into consciousness and being remembered, if initially again in the *condition seconde*, but that they were a burden and disturbance to the normal state of consciousness. It remains to be seen whether the same origin can be attributed to other cases of chronic hysteria which conclude with a psychosis.[8]

Before beginning experiments that gave life to his evil alter Edward Hyde, Dr. Jekyll had intimations of a "profound duplicity" within himself.

ROBERT LOUIS STEVENSON, *THE STRANGE CASE OF DR. JEKYLL AND MR. HYDE* (1886)

With every day, and from both sides of my intelligence, the moral and the intellectual, I thus drew steadily nearer to that truth, by whose partial discovery I have been doomed to such a dreadful shipwreck: that man is not truly one, but truly two. I say two, because the state of my own knowledge does not pass beyond that point. Others will follow, others will outstrip me on the same lines; and I hazard the guess that man will be ultimately known for a mere polity of multifarious, incongruous and independent denizens. I for my part, from the nature of my life, advanced infallibly in one direction and in one direction only. It was on the moral side, and in my own person, that I learned to recognize the thorough and primitive duality of man; I saw that, of the two natures that contended in the field of my consciousness, even if I could rightly be said to be either, it was only because I was radically both; and from an early date, even before the course of my scientific discoveries had begun to suggest the most naked possibility of such a miracle, I had learned to dwell with plea-sure, as a beloved daydream, on the thought of the separation of these elements. If each, I told myself, could but be housed in separate identi-ties, life would be relieved of all that was unbearable; the unjust might go his way, delivered from the aspirations and remorse of his more upright twin; and the just could walk steadfastly and securely on his upward path, doing the good things in which he found his pleasure, and no longer exposed to disgrace and penitence by the hands of this extraneous evil. It was the curse of mankind that these incongruous faggots were thus bound together—that in the agonized womb of consciousness, these polar twins should be continuously struggling. How, then, were they dissociated?[9]

W. E. B. Du Bois (1868–1963), one-time student of William James, used the concept of double consciousness to describe the inner conflict of his fellow Black Americans.

W. E. B. DU BOIS, *THE SOULS OF BLACK FOLK* (1903)

It is a peculiar sensation, this double-consciousness, this sense of always looking at one's self through the eyes of others, of measuring one's soul by the tape of a world that looks on in amused contempt and pity. One ever feels his two-ness,—an American, a Negro; two souls, two thoughts, two unreconciled strivings; two warring ideals in one dark body, whose dogged strength alone keeps it from being torn asunder.

The history of the American Negro is the history of this strife,—this longing to attain self-conscious manhood, to merge his double self into a better and truer self. In this merging he wishes neither of the older selves to be lost. He would not Africanize America, for America has too much to teach the world and Africa. He would not bleach his Negro soul in a flood of white Americanism, for he knows that Negro blood has a message for the world. He simply wishes to make it possible for a man to be both a Negro and an American, without being cursed and spit upon by his fellows, without having the doors of Opportunity closed roughly in his face.[10]

The Swiss psychiatrist Eugen Bleuler (1857–1939) coined the terms schizophrenia and depth psychology. Here, Bleuler reflects on the oddities of the "double-entry bookkeeping" he observed on in the psychiatric wards of Burghölzli Hospital.

EUGEN BLEULER, *DEMENTIA PRAECOX* (1911)

The splitting of the psyche into several souls always leads to the greatest inconsistencies. A persecuted patient, upon release from the institution, took leave, movingly and with real emotion, of her chief tormentor who had so often wanted to kill her. The patients will confidently hand us their letters to be forwarded, in which they accuse us of the most atrocious crimes, as well as of constantly defrauding them of their mail. They curse us in the strongest terms as their poisoners, only to ask us in the very next moment to examine them for some minor ailment, or to ask for a cigarette. . . .

It is especially important to know that these patients carry on a kind of "double-entry bookkeeping" in many of their relationships. They know the real state of affairs as well as the falsified one and will answer according to the circumstances with one kind or the other type of orientation—or both together. This last is especially frequent in misrecognizing people: the physician "is now here as Dr. N.," at other times he becomes the former lover.[11]

Following the demise of hysteria as a disease category, dissociation or "splitting-off" remained a fundamental concept in psychiatry, particularly in relation to dementia praecox (schizophrenia).

KARL JASPERS, *GENERAL PSYCHOPATHOLOGY* (1913)

Radical dissociation (splitting-off) is abnormal in every case and so is its inaccessibility for consciousness, its failure to integrate into the personality and the disruption of continuity with the individual life as a whole. This dissociation (splitting-off) is to be *sharply demarcated* from those divisions of normal life that commonly reunite again into context. Dissociation (splitting-off)—like the crossing of a Rubicon demarcates anarchy from unified experience. *Interpretation according to the category of dissociation* occurs with numerous modifications: neurotic symptoms, organ-complaints, come to be regarded as phenomena torn away from their meaningful life-source. Independence of apparatus leads, for example, to uninhibited isolation of the sensory fields. The term dissociation (splitting-off) is given to the inability to remember experiences which remain effective none the less. A lack of relationship in psychic development, the disintegration of integrated wholes, unconnected double-meanings, double-interpretations and similar phenomena in dementia praecox have led to the use of the term dissociation insanity (schizophrenia). Experiences of a double-self are called dissociation of the self. The continual problem is *what is it* that produces this *tearing apart* and *by what means* can *re-integration* take place and with it the restoration of meaning, definition and proportion.[12]

Though fin-de-siècle psychiatry maintained that true dissociation was always pathological, psychologists and psychical researchers continued to be enthralled by mediums and automatists whose secondary personalities eclipsed their everyday selves.

CHARLES E. CORY, "PATIENCE WORTH" (1919)

Patience Worth, the writer, is a subconscious personality of Mrs. John Curran, of St. Louis. About five years ago Mrs. Curran began to write, automatically, literature of an unusual character. Since that time novels, plays, and poems have appeared. Over fifteen hundred poems have been written. Two of the novels, 'The Sorry Tale' and 'Hope Trueblood,' have been published by Henry Holt & Co. Four additional novels are in various stages of completion. Most of this literature is conceded by critics to be of a very high order. . . . Mrs. Curran is an intelligent woman, but her mind is much inferior to that of Patience Worth. In short, here is a subconscious self far outstripping in power and range the primary consciousness. This is an indisputable fact, and it is a significant one for psychology. In some way the dissociation has resulted in the formation of a self with greatly increased caliber. It has not only given it access to a much wider range of material, but it has given it a facile creative power amounting to genius. . . .

It is evident that the term subconscious is misleading when used to describe the source of this literature. As generally used it would imply that these works are the product of marginal or submarginal tendencies. This they are, only with reference to the other field. With reference to the self that created them, they are distinctly within the conscious. The term co-conscious, as used by [Walter Franklin] Prince to describe somewhat similar cases, is helpful here. At all events, and this is the significant thing, these delicate and finely rational processes, these highly elaborate compositions, are performed apparently without the aid or knowledge of Mrs. Curran.[13]

In the last of his major theoretical works, Freud retrofitted the internal dynamics of the unconscious, describing the internecine power-play of the ego, id, and superego.

SIGMUND FREUD, "THE EGO AND THE ID" (1923)

The ego represents what may be called reason and common sense, in contrast to the id, which contains the passions. All this falls into line with popular distinctions which we are all familiar with; at the same time, however, it is only to be regarded as holding good on the average or 'ideally'. The functional importance of the ego is manifested in the fact that normally control over the approaches to motility devolves upon it. Thus in its relation to the id it is like a man on horseback, who has to hold in check the superior strength of the horse; with this difference, that the rider tries to do so with his own strength while the ego uses borrowed forces. The analogy may be carried a little further. Often a rider, if he is not to be parted from his horse, is obliged to guide it where it wants to go; so in the same way the ego is in the habit of transforming the id's will into action as if it were its own.[14]

In the founding text of ego psychology, Anna Freud (1895–1982) amplified the ego's capacity to control and regulate the id's instincts and impulses through various defensive ruses.

ANNA FREUD, *THE EGO AND THE MECHANISMS OF DEFENCE* (1936)

In the id the so-called "primary process" prevails; there is no synthesis of ideas, affects are liable to displacement, opposites are not mutually exclusive and may even coincide, and condensation occurs as a matter of course. The sovereign principle which governs the psychic processes is that of obtaining pleasure. In the ego, on the contrary, the association of ideas is subject to strict conditions, to which we apply the comprehensive term "secondary process"; further, the instinctual impulses can no longer seek direct gratification—they are required to respect the demands of reality and, more than that, to conform to ethical and moral laws by which the superego seeks to control the behavior of the ego. Hence these impulses run the risk of incurring the displeasure of institutions essentially alien to them. They are exposed to criticism and rejection and have to submit to every kind of modification. Peaceful relations between the neighboring powers are at an end. The instinctual impulses continue to pursue their aims with their own peculiar tenacity and energy, and they make hostile incursions into the ego, in the hope of overthrowing it by a surprise attack. The ego on its side becomes suspicious; it proceeds to counterattack and to invade the territory of the id. Its purpose is to put the instincts permanently out of action by means of appropriate defensive measures, designed to secure its own boundaries.[15]

While Freudians regarded schizophrenia as beyond the therapeutic reach of psychoanalysis, Melanie Klein (1882–1960) and the Object Relations School found that paranoid defenses and schizoid ambivalence could be breached through transference.

MELANIE KLEIN, *NOTES ON SOME SCHIZOID MECHANISMS* (1946)

It is generally agreed that schizoid patients are more difficult to analyse than manic-depressive types. Their withdrawn, unemotional attitude, the narcissistic elements in their object-relations (to which I referred earlier), a kind of detached hostility which pervades the whole relation to the analyst create a very difficult type of resistance. I believe that it is largely the splitting processes which account for the patient's failure in contact with the analyst and for his lack of response to the analyst's interpretations. The patient himself feels estranged and far away, and this feeling corresponds to the analyst's impression that considerable parts of the patient's personality and of his emotions are not available. Patients with schizoid features may say: 'I hear what you are saying. You may be right, but it has no meaning for me.' Or again they say they feel they are not there. The expression 'no meaning' in such cases does not imply an active rejection of the interpretation but suggests that parts of the personality and of the emotions are split off. These patients can, therefore, not deal with the interpretation; they can neither accept it nor reject it.

I shall illustrate the processes underlying such states by a piece of material taken from the analysis of a man patient. The session I have in mind started with the patient's telling me that he felt anxiety and did not know why. He then made comparisons with people more successful and fortunate than himself. These remarks also had a reference to me. Very strong feelings of frustration, envy and grievance came to the fore. When I interpreted—to give here again only the gist of my interpretations—that these feelings were directed against the analyst and that he wanted to destroy me, his mood changed abruptly. The tone of his voice became flat, he spoke in a slow, expressionless way, and he said that he felt detached from the whole situation. He added that my interpretation seemed correct, but that it did not matter. In fact, he no longer had any wishes, and nothing was worth bothering about.

My next interpretations centred on the causes for this change of mood. I suggested that at the moment of my interpretation the danger of destroying me had become very real to him and the immediate consequence was the fear of losing me. Instead of feeling guilt and depression, which at certain stages of his analysis followed such interpretations, he now attempted to deal with these dangers by a particular method of splitting. As we know, under the pressure of ambivalence, conflict and guilt, the patient often splits the figure of the analyst; then the analyst may at certain moments be loved, at other moments hated. Or the relations to the analyst may be split in such a way that he remains the good (or bad) figure while somebody else becomes the opposite figure. But this was not the kind of splitting which occurred in this particular instance. The patient split off those parts of himself, i.e. of his ego which he felt to be dangerous and hostile towards the analyst. He turned his destructive impulses from his object towards his ego, with the result that parts of his ego temporarily went out of existence. In unconscious phantasy this amounted to annihilation of part of his personality. The particular mechanism of turning the destructive impulse against one part of his personality, and the ensuing dispersal of emotions, kept his anxiety in a latent state.[16]

George Orwell (1903–1950) foresees the instrumental use of dissociation.

GEORGE ORWELL, *NINETEEN EIGHTY-FOUR* (1948)

All that was needed was an unending series of victories over your own memory. 'Reality control', they called it: in Newspeak, 'doublethink.' . . . To know and not to know, to be conscious of complete truthfulness while telling carefully constructed lies, to hold simultaneously two opinions which cancelled out, knowing them to be contradictory and believing in both of them, to use logic against logic, to repudiate morality while laying claim to it, to believe that democracy was impossible and that the Party was the guardian of democracy, to forget whatever it was necessary to forget, then to draw it back into memory again at the moment when it was needed, and then promptly to forget it again: and above all, to apply the same process to the process itself. That was the ultimate subtlety:

consciously to induce unconsciousness, and then, once again, to become unconscious of the act of hypnosis you had just performed. Even to understand the word 'doublethink' involved the use of doublethink.[17]

Might auditory hallucinations, a first-rank symptom of schizophrenia, be generated by subvocal speech?

LOUIS N. GOULD, "VERBAL HALLUCINATIONS AS AUTOMATIC SPEECH" (1950)

The identity of automatic speech to verbal hallucinations was unquestionably shown [by auscultation of the larynx and electromyography of the vocal organs] in the case of the patient L. M. She exhibited almost continuous, involuntary, faint whispering accompanied by movements of the floor of the mouth. Upon electronic amplification there was found marked correspondence in content and character of the subaudible speech to the "voices" as reported by the subject. Automatic speech from time to time varied in distinctness and intensity and was different in emphasis, quality, rhythm, and rate (being about twice as fast) from her voluntary whisper and speech. The inspiratory and expiratory phases of respiration employed in speech production corresponded to the two voices she stated she was hearing. Proprioceptive sensations from the vocal organs were attributed by the patient to the verbal hallucinations. Upon administration of sodium amytal followed by caffeine sodium benzoate, both intravenously, loud whispering at the normal rate was followed by faint whispering at twice the rate. The latter was again interpreted by the patient as voices. . . .

That the verbal hallucination is automatic speech has been proved . . . Patients have admitted that they were talking to themselves or were hearing their own thoughts. Some identified thinking with talking. Some experienced voices issuing from their mouths or throat; many felt proprioceptive sensations from involuntary activity of larynx, mouth, tongue, or lips accompanying the verbal hallucination or the thought. A few patients reported that this activity interfered with voluntary speech.

There is also objective proof of automatic speech. Involuntary movement of organs of speech and in one patient ventriloquist-like behavior have been observed by others. Misinterpretation as voices by a patient of her automatic, almost continuous faint whispering was demonstrated by the author. Other nonverbal oral sounds produced automatically were also interpreted as foreign.[18]

R. D. Laing's (1927–1989) patient "Rose" exemplified what the Scottish psychiatrist found to be the most elementary defense in every psychosis: "the denial of being, as a means of preserving being."

R. D. LAING, *THE DIVIDED SELF* (1960)

[S]he felt she could not reach other people, that other people could not reach her, and the more she felt herself to be in a world of her own—'They can't get in and I can't get out'—the more this private and closed world of hers became invaded by psychotic dangers from outside, i.e. the more 'public' in a sense it became. She became more suspicious of other people and began to hide things in her locker; she had a notion that some person was stealing things from her. She would check her handbag and personal possessions frequently to make sure that she had not been robbed of anything. This paradox of being more withdrawn and at the same time more vulnerable found its clearest expression in the statement that she was murdering herself on the one hand, and her fear that her 'self' might be lost or stolen on the other. She had only other people's thoughts and could think only what other people had said.

She now talked of being two. 'There are two mes.' 'She's me, and I'm her all the time.' She heard a voice telling her to murder her mother and she knew that this voice belonged to 'one of my mes'. 'From up here [indicating her temples] it's just cotton wool. I've no thoughts of my own; I'm awfully confused, me, me, me all the time, me and me, me and myself, when I say myself, I know there's something wrong, something's happening to me, I don't know what.'

Thus, despite the fear of losing her self, all her efforts to 'recapture reality' involved not being herself, and attempts to escape from her self or to kill her self continued to be used as basic defences, indeed, they became intensified.[19]

The French poet Henri Michaux (1899–1984) on the critical "other"
experienced while writing under the effects of mescaline.

HENRI MICHAUX, *THE MAJOR ORDEALS OF THE MIND* (1966)

What he cannot prevent is a marginal and continual intervention, active
and critical, an insistent presence which does not let a single word pass
without interfering, which magisterially parts the words, like a pair of
doors, to insert itself among them, to introduce its own reflections, the
reflection of a witness concerned with everything and generally uncoop-
erative—and there is nothing he can do to prevent it. This presence also
keeps moving, has its own powerful, unsuspected movements, advances,
retreats, returns like someone really there, close enough to touch him.

The writing continues in this fashion, "supervised" by the other. Not
only by one. It is now a sort of murmur, as though from a group of several
who might intercede, who intercede among the words, between one word
and another, between one idea and its opposite, and interrupt, and interfere, and grumble and object, and mock, and disapprove, and jeer, and
say "perhaps" and "perhaps not" and "not at all," and reconsider, and
do not tolerate, and argue, and dissent, and laugh, and laugh, and laugh,
and hop around, and clatter about, keep clattering, meanly, increasingly,
continually, incredibly.[20]

The British psychoanalyst Donald Winnicott (1896–1971) found that, for
some patients, dissociation was also an elective escape from reality

DONALD WINNICOTT, *PLAYING AND REALITY* (1971)

All the time, without her knowing it, while she was at school and later at
work, there was another life going on in terms of the part that was dissociated. Put the other way around, this meant that her life was dissociated
from the main part of her, which was living in what became an organized
sequence of fantasying. If one were to trace this patient's life one could
see the ways in which she attempted to bring together these two and
other parts of her personality, but her attempts always had some kind of
protest in them which brought a clash with society. All the time she had

enough health to continue to give promise and to make her relations and
her friends feel that she would make her mark, or at any rate that she
would one day enjoy herself. To fulfil this promise was impossible, however, because (as she and I have gradually and painfully discovered) the
main part of her existence was taking place when she was doing nothing
whatever. . . . As soon as this patient began to put something into practice, such as to paint or to read, she found the limitations that made her
dissatisfied because she had let go of the omnipotence that she retained
in the fantasying. This could be referred to in terms of the reality principle but it is more true, in the case of a patient like this, to speak of the
dissociation that was a fact in her personality structure.[21]

*The French theorists Gilles Deleuze (1925–1995) and Felix Guattari (1930–
1992) turned schizophrenia's descent into the unconscious into a conceptual
romance of living at "the very limits of the decoded flows of desire."
Breakthrough, not breakdown.*

GILLES DELEUZE AND FELIX GUATTARI, *ANTI-OEDIPUS: CAPITALISM AND SCHIZOPHRENIA* (1972)

Our society produces schizos the same way it produces Prell shampoo or
Ford cars, the only difference being that the schizos are not salable. How
then does one explain the fact that capitalist production is constantly
arresting the schizophrenic process and transforming the subject of the
process into a confined clinical entity, as though it saw in this process
the image of its own death coming from within? Why does it make the
schizophrenic into a sick person not only nominally but in reality? Why
does it confine its madmen and madwomen instead of seeing in them its
own heroes and heroines, its own fulfillment? And where it can no longer
recognize the figure of a simple illness, why does it keep its artists and
even its scientists under such close surveillance—as though they risked
unleashing flows that would be dangerous for capitalist production and
charged with a revolutionary potential, so long as these flows are not co-
opted or absorbed by the laws of the market? Why does it form in turn
a gigantic machine for social repression—psychic repression, aimed at
what nevertheless constitutes its own reality—the decoded flows? The

answer—as we have seen—is that capitalism is indeed the limit of all societies, insofar as it brings about the decoding of the flows that the other social formations coded and overcoded. . . . Schizophrenia, on the contrary, is indeed the absolute limit that causes the flows to travel in a free state on a desocialized body without organs. Hence one can say that schizophrenia is the exterior limit of capitalism itself or the conclusion of its deepest tendency.[22]

Princeton psychologist Julian Jaynes (1920–1997) proposed that the unified consciousness is approximately 3,000 years old. Before this, our forebears would have had limited capacity for introspection, the true agents of their affairs being the voices of gods.

JULIAN JAYNES, *THE ORIGIN OF CONSCIOUSNESS IN THE BREAKDOWN OF THE BICAMERAL MIND* (1977)

Who then were these gods that pushed men about like robots and sang epics through their lips? They were voices whose speech and directions could be as distinctly heard by the Iliadic heroes as voices are heard by certain epileptic and schizophrenic patients, or just as Joan of Arc heard her voices. The gods were organizations of the central nervous system and can be regarded as personae in the sense of poignant consistencies through time, amalgams of parental or admonitory images. The god is a part of the man, and quite consistent with this conception is the fact that the gods never step outside of natural laws. Greek gods cannot create anything out of nothing, unlike the Hebrew god of Genesis. In the relationship between the god and the hero in their dialectic, there are the same courtesies, emotions, persuasions as might occur between two people. The Greek god never steps forth in thunder, never begets awe or fear in the hero, and is as far from the outrageously pompous god of Job as it is possible to be. He simply leads, advises, and orders. Nor does the god occasion humility or even love, and little gratitude. Indeed, I suggest that the god-hero relationship was—by being its progenitor—similar to the referent of the ego-superego relationship of Freud or the self-generalized other relationship of Mead. The strongest emotion which the hero feels toward a god is amazement or wonder, the kind of emotion that we feel when the solution of a particularly difficult problem suddenly pops into our heads, or in the cry of eureka! from Archimedes in his bath.[23]

In the wake of Roger Sperry's landmark split-brain experiments at the California Institute of Technology, neuroscientists began to consider specialized brain hemispheres as akin to separate selves.

JOSEPH LEDOUX, DONALD H. WILSON, AND MICHAEL S. GAZZANIGA, "A DIVIDED MIND" (1977)

The question of whether the essence of human consciousness can be represented bilaterally in the split-brain patient has so far remained unanswered. The following observations on a new patient, Patient P. S., may help to resolve the issue. For the first time, it has been possible to ask subjective questions of the separated right hemisphere and to witness self-generated answers from this mute half-brain. This opportunity was made possible by the fact that linguistic representation in the right hemisphere of our patient is greater than has been observed in any other split-brain patient. In addition to an extensive capacity for comprehending written and spoken language, the right hemisphere, though unable to generate speech, can express its mental content by arranging letters to spell words. . . .

The right half-brain spelled "Paul" in response to the question "Who are you!" When requested to spell his favorite girl, the right hemisphere arranged the Scrabble letters to spell "Liz." The right hemisphere spelled "car" for his favorite hobby. When the right hemisphere was asked to spell his favorite person, the following was generated: "Henry Wi Fozi." (Henry Winkler is the actor who plays Fonzie.) The right hemisphere generated "Sunday" in response to the question "What is tomorrow?" When asked to describe his mood, the right hemisphere spelled out "good." Later, in response to the same question, the left spelled "silly." Finally, the right hemisphere spelled out "automobile race" as the job he would pick. This contrasts with the frequent assertion of the left hemisphere that he will be a "draftsman." In fact, shortly after the test session, when asked what he would like to do for a living, the left hemisphere said, "Oh, be a draftsman, I guess." Although hand use was not dictated, the left hand dominated the spelling responses of the right hemisphere, receiving some occasional support from the right hand. Finally, it should be noted that on each of these right hemisphere trials the patient was unable to name the lateralized information, thus confirming that the left hemisphere did not have access to the critical information. . . .

Since the conscious properties of the left hemisphere are obvious through a subject's verbal behavior, our main concern has been with the silent inhabitant of the right side of the cranium. In this regard, we have found that the right hemisphere in this patient has a sense of self, for it knows the name it collectively shares with the left. The right hemisphere has feelings, for it can describe its mood. The right hemisphere has a sense of who it likes and what it likes to do. The right hemisphere has a sense of the future, for it knows what day tomorrow is. The right hemisphere has aspirations and goals for the future, for it can describe its occupational choice. . . .

This clear example of surgically produced psychological dynamism, seen for the first time in P. S., raises the question whether such processes are active in the normal brain, where different mental systems, using different neural codes, coexist within and between the cerebral hemispheres.[24]

Philip K. Dick's (1928–1982) science fiction classic A Scanner Darkly, *in which narcotics agent Bob Arctor (a.k.a Fred) succumbs to the divisory effects of Substance-D, drew heavily on findings from split-brain research.*

PHILIP K. DICK, *A SCANNER DARKLY* (1977)

Back at Room 203, the police psychology testing lab, Fred listened without interest as his test results were explained to him by both the psychologists.

"You show what we regard more as a competition phenomenon than impairment. Sit down."

"Okay," Fred said stoically, sitting down.

"Competition," the other psychologist said, "between the left and right hemispheres of your brain. It's not so much a single signal, defective or contaminated; it's more like two signals that interfere with each other by carrying conflicting information."

"Normally," the other psychologist explained, "a person uses the left hemisphere. The self-system or ego, or consciousness, is located there. It is dominant, because it's in the left hemisphere always that the speech center is located; more precisely, bilateralization involves a verbal ability on valency in the left, with spatial abilities in the right. The left can be compared to a digital computer; the right to an analogic. So bilateral function is not mere duplication; both percept systems monitor and process incoming data differently. But for you, neither hemisphere is dominant and they do not act in a compensatory fashion, each to the other. One tells you one thing, the other another." . . .

"I'm sure you know already," the psychologist to the left said. "You've been experiencing it, without knowing why or what it is."

"The two hemispheres of my brain are competing?" Fred said.

"Yes."

"Why?"

"Substance D. It often causes that, functionally. This is what we expected; this is what the tests confirm. Damage having taken place in the normally dominant left hemisphere, the right hemisphere is attempting to compensate for the impairment. But the twin functions do not fuse, because this is an abnormal condition the body isn't prepared for. It should never happen. *Cross-cuing*, we call it. Related to split-brain phenomena."[25]

As hysteria lost its place as a disease category, Multiple Personality Disorder filled the void.

AMERICAN PSYCHIATRIC ASSOCIATION, *DIAGNOSTIC AND STATISTICAL MANUAL OF MENTAL DISORDERS*, THIRD EDITION (1980)

The essential feature is the existence within the individual of two or more distinct personalities, each of which is dominant at a particular time. Each personality is a fully integrated and complex unit with unique memories, behavior patterns, and social relationships that determine the nature of the individual's acts when that personality is dominant. Transition from one personality to another is sudden and often associated with psychosocial stress. Usually the original personality has no knowledge or awareness of the existence of any of the other personalities (subpersonalities). When there are more than two subpersonalities in one individual, each is aware of the others to varying degrees. The subpersonalities may not know each other or be constant companions. At any given moment one personality will interact verbally with the external environment, but none or any number of the other personalities may actively perceive (i.e., "listen in on") all that is going on. The original personality and all of the subpersonalities are aware of lost periods of time. "They" will usually admit to this if asked, but will seldom volunteer this information. The individual personalities are nearly always quite discrepant and frequently seem to be opposites. For example, a quiet, retiring spinster may alternate with a flamboyant, promiscuous bar habitue on certain nights. Usually one of the personalities over the course of the disorder is dominant. Associated features. One or more of the personalities may function with a reasonable degree of adaptation (e.g., be gainfully employed) while alternating with another personality that is clearly maladapted or has a specific, separate, mental disorder. Studies have demonstrated that various subpersonalities in the same individual may have different responses to physiological and psychological measurements.[26]

Were psychiatrists aiding and abetting the creation of a factitious disorder? Corbett Thigpen (1919–1999) and Hervey Cleckley (1903–1984), co-authors of the 1957 multiple-personality classic The Three Faces of Eve, *raised an early alarm.*

CORBETT H. THIGPEN AND HERVEY M. CLECKLEY, "ON THE INCIDENCE OF MULTIPLE PERSONALITY DISORDER" (1984)

The severe dissociation characteristic of multiple personality disorder is most often considered to result from the patient's inability to cope in other ways with intense conflicting impulses or feelings. Before assuming, however, that the patient's "personalities" or "ego fragments" are long-standing, autonomous entities that are fully dissociated from the original personality—and therefore serve separate motives, impulses, and feelings—we have found it useful to consider whether there might not be instead a pseudo- or quasi-dissociation that functions to help the patient gain attention, or maintain an acceptable self-image, or accrue financial gain, or even escape responsibility for actions. This often turns out to be the case, and while the patient may consciously or unconsciously desire to achieve a total dissociation (i.e., to be diagnosed as having multiple personality) as a way out of a difficult psychological or emotional dilemma, this does not mean that the individual either truly experiences the degree of dissociation necessary for a diagnosis of multiple personality, or remains amnesic for other "parts" of himself or herself.

Nevertheless, there seems to be a tendency these days to ignore or underestimate secondary gain as a motivation for manifesting quasi- or pseudo-dissociative symptoms by individuals. This has become, in our view, a major problem with a patient who has been charged with criminal activity. There have recently been two celebrated legal cases (one involving rape and the other rape/murder) where the defendants had been diagnosed as having a multiple personality, and pleaded not guilty by reason of insanity. Clearly, there is a good deal for the individual to gain by being diagnosed as having multiple personality disorder. Indeed, in one of these cases (State v. Milligan, 1978), the diagnosis was largely unchallenged and the defendant's plea was upheld.[27]

8

MIND AT LARGE: NEW AGE SCIENCE AND RELIGION

In 1918, the German sociologist Max Weber famously observed that: "The fate of our age is characterized by rationalization and intellectualization and, above all, by the disenchantment of the world."[1] Weber's lament on the demise of the sacred and supernatural came with an often-ignored caveat. The "iron cage" of modernity could not, he added, contain or suppress the inner needs of the human beings who had religion in their veins. Science and secularism would, Weber predicted, be haunted by novel forms of magic, wonder, and mystery.

The counter-revolution that Weber prophesied was, of course, already stirring. On both sides of the Atlantic, spiritism, occultism, and theosophy remained active crucibles of faith in the supernatural for freethinkers and radicals; psychical research, taking leave of the seance-room, was continuing to shape the field of subliminal psychology; and Christian Science, having thoroughly plundered the lessons of mesmerism, continued to minister its own metaphysics of health and wellbeing.

Over the coming century, the exploration and expansion of the unconscious became a central pillar of the New Age movement. Psychedelic visions, the meditative awakenings of samadhi, satori, kensho, peak, or right-brain experiences, gestaltism, UFO abductions, near-death experiences, past-life regressions, visions of angels—these were stations on the via mystica of mind expansion, sideshows in the psychospiritual carnival that rolled out of the foothills of central California.

Significantly, the unconscious as championed by the prophets of self-transcendence and self-actualization was for the most part antithetical to Freud, who had, after all, done no more than pledge to replace the misery of neurosis with "ordinary unhappiness." Neither personal nor flesh-bound, the New Age unconscious promised an enlargement of being, a tipping of the self into a boundless "cosmic consciousness."

Central to the growing preoccupation with inner growth and "mind expansion" was the translation and popularization of key texts from the Vedanta School of Hinduism, Tibetan Buddhism, Japanese Zen, and Chinese Taoism, whose spiritual technologies appeared ripe with secrets that Western psychology had been slow to grasp. And of growing importance to the emergent counterculture were the "techniques of ecstasy" that Mircea Eliade described in his pioneering book *Shamanism*.

The proliferation of experiential pathways promising access to higher states of consciousness was greeted with skepticism in many quarters. The New Age's syncretic pick 'n' mix approach to religiosity did not sit well with theologians who scoffed at its counterfeit mysticism, or with Left-leaning theorists who, with some justification, accused the counterculture of forsaking political engagement for collective solipsism. Both camps were, following Harold Bloom, literary critic turned New Age sleuth, well off the mark. For Bloom, angelic visions, psychic experiences, and near-death experiences belonged to a bona fide tradition of American hermeticism, itself an offshoot of a grand gnostic tradition of inner exploration that has shadowed every major religion.

If this quest for the deep self is the defining feature of New Age religiosity, its psychological keynotes do nevertheless differ in some important respects. Encounters with the unconscious sometimes involve an escape from linear time, a dissolution of the ego, a union of the masculine and feminine, an experience of oneness and non-duality, communication outside of language, passage into a numinous, sacred or cosmic dimension that may involve contact with other beings.

Within this spiritualization of depth psychology, there is, of course, another more mundane matter to consider: commerce. Purveyors of biofeedback gadgetry, lucid dream machines, right-brain training modules, or courses in human potential are rarely shy of bluster and hyperbole. In fact, even the sincerest navigators of the unconscious mind have, as we will now see, a tendency to espouse credos that treat the unconscious as a deus ex machina. To use a term on which Weber leant when describing the relationship between Protestantism and capitalism, one might even say that opportunism and the idea of the unconscious have always enjoyed an "elective affinity."

Vedantic philosophy, the cornerstone of Hinduism, was a significant influence on American Transcendentalism, and Ralph Waldo Emerson's (1803–1882) concept of a divine over-soul chimed with the supreme self or Paramatman of the Upanishads.

RALPH WALDO EMERSON, *THE OVER-SOUL* (1841)

The philosophy of six thousand years has not searched the chambers and magazines of the soul. In its experiments there has always remained, in the last analysis, a residuum it could not resolve. Man is a stream whose source is hidden. Our being is descending into us from we know not whence. The most exact calculator has no prescience that somewhat incalculable may not balk the very next moment. I am constrained every moment to acknowledge a higher origin for events than the will I call mine.

As with events, so is it with thoughts. When I watch that flowing river, which, out of regions I see not, pours for a season its streams into me, I see that I am a pensioner; not a cause, but a surprised spectator of this ethereal water; that I desire and look up, and put myself in the attitude of reception, but from some alien energy the visions come.

The Supreme Critic on the errors of the past and the present, and the only prophet of that which must be, is that great nature in which we rest, as the earth lies in the soft arms of the atmosphere; that Unity, that Over-soul, within which every man's particular being is contained and made one with all other; that common heart, of which all sincere conversation is the worship, to which all right action is submission; that overpowering reality which confutes our tricks and talents, and constrains every one to pass for what he is, and to speak from his character, and not from his tongue, and which evermore tends to pass into our thought and hand, and become wisdom, and virtue, and power, and beauty. We live in succession, in division, in parts, in particles. Meantime within man is the soul of the whole; the wise silence; the universal beauty, to which every part and particle is equally related; the eternal ONE. And this deep power in which we exist, and whose beatitude is all accessible to us, is not only self-sufficing and perfect in every hour, but the act of seeing and the thing seen, the seer and the spectacle, the subject and the object, are one. We see the world piece by piece, as the sun, the moon, the animal, the tree; but the whole, of which these are the shining parts, is the soul. Only

by the vision of that Wisdom can the horoscope of the ages be read, and
by falling back on our better thoughts, by yielding to the spirit of prophecy which is innate in every man, we can know what it saith. Every man's
words, who speaks from that life, must sound vain to those who do not
dwell in the same thought on their own part. I dare not speak for it. My
words do not carry its august sense; they fall short and cold. Only itself
can inspire whom it will, and behold! their speech shall be lyrical, and
sweet, and universal as the rising of the wind. Yet I desire, even by profane words, if I may not use sacred, to indicate the heaven of this deity,
and to report what hints I have collected of the transcendent simplicity
and energy of the Highest Law.[2]

The Easternization of the Yankee unconscious.

ROBERT C. FULLER, *AMERICANS AND THE UNCONSCIOUS* (1995)

From Emerson on, the aesthetic strain in American religious thought has
often called upon the metaphorical and symbolic language of Eastern
mystical traditions. The Hindu monistic philosophy, which establishes
a fundamental unity between the Atman (individual consciousness) and
Brahman (the Over-Soul or world consciousness), supplied both Emerson and Thoreau with the psychological underpinnings of self-reliance.
New Thoughters such as Warren Felt Evans also used Eastern terminology
to explain the individual's connection with creative spiritual energies.
Their writings extol Eastern metaphysical traditions for nurturing belief
in the psyche's spiritual depths in ways that the Christian tradition does
not. The pantheistic theologies and meditative psychologies of the Orient
supply a ready-made vocabulary for expressing the harmonial conviction
that inward journeys lead to higher, spiritual worlds.

American appropriation of Eastern metaphors for the successive layers
of the unconscious mind is perhaps not as meaningful as it seems at first
glance. Most American adherents of Eastern traditions actually know pre-
cious little about the cultural or historical origins of the symbols they so
avidly embrace. American versions of Zen, Theosophy, Vedanta, and Yoga
are little more than a repackaging of indigenous spirituality. Individuals
whose aesthetic rather than doctrinal approach to religion prompts them

to sever ties with the Judeo-Christian scriptural tradition find in Eastern religions a ready-made vocabulary for articulating their profoundly incarnational and sacramental view of the universe. In Eastern teachings they find reassurance that the deepest realms of their own unconscious minds provide a portal through which a transcendent spiritual power can enter into and illuminate their lives.

Almost nowhere in American versions of Eastern metaphysics does one find ideas that contradict the substance or style of Yankee optimism.[3]

The occult psychology of Helena Blavatsky (1831–1891), the grande dame of Theosophy, proposed the existence of multiple layers of consciousness, all under the awning of a vast "universal consciousness."

HELENA BLAVATSKY, *THE SECRET DOCTRINE* (1888)

"Mind" is a name given to the sum of the States of Consciousness, grouped under Thought, Will and Feeling. During deep sleep ideation ceases on the physical plane, and memory is in abeyance; thus for the time-being "Mind is not," because the organ, through which the Ego manifests ideation and memory on the material plane, has temporarily ceased to function. A noumenon can become a phenomenon on any plane of existence only by manifesting on that plane through an appropriate basis or vehicle; and during the long Night of rest called Pralaya, when all the Existences are dissolved, the "Universal Mind" remains as a permanent possibility of mental action, or as that abstract absolute Thought, of which Mind is the concrete relative manifestation. The Ah-hi (Dhyân Chohans) are the collective hosts of spiritual Beings—the Angelic Hosts of Christianity, the Elohim and "Messengers" of the Jews—who are the Vehicle for the manifestation of the Divine or Universal Thought and Will. They are the Intelligent Forces that give to, and enact in, Nature her "Laws," while they themselves act according to Laws imposed upon them in a similar manner by still higher Powers; but they are not the "personifications" of the Powers of Nature, as erroneously thought. This Hierarchy of spiritual Beings, through which the Universal Mind comes into action, is like an army—a host, truly—by means of which the fighting power of a nation manifests itself, and which is composed of army-corps, divisions, brigades, regiments, and so forth, each with its separate individuality or life, and its limited freedom of action and limited responsibilities; each contained in a larger individuality, to which its own interests are subservient, and each containing lesser individualities in itself.[4]

Pioneering reformer and socialist Edward Carpenter (1844–1929) was among the vanguard of radicals who embraced the liberatory promise of cosmic consciousness emanating from the East.

EDWARD CARPENTER, "A CONSCIOUSNESS WITHOUT THOUGHT" (1892)

Great have been the disputes among the learned as to the meaning of the word *Nirvâna*—whether it indicates a state of no-consciousness or a state of vastly enhanced consciousness. Probably both views have their justification; the thing does not admit of definition in the terms of ordinary language. The important thing to see and admit is that under cover of this and other similar terms there does exist a real and recognizable fact (that is, a state of consciousness in some sense), which has been experienced over and over again, and which to those who have experienced it in ever so slight a degree has appeared worthy of lifelong pursuit and devotion. It is easy of course to represent the thing as a mere word, a theory, a speculation of the dreamy Hindu; but people do not sacrifice their lives for empty words, nor do mere philosophical abstractions rule the destinies of continents. No, the word represents a reality, something very basic and inevitable in human nature. The question really is not to define the fact—for we cannot do that—but to get at and experience it.

It is interesting at this juncture to find that modern Western science, which has hitherto—without much result—been occupying itself with mechanical theories of the universe, is approaching from its side this idea of the existence of another form of consciousness. The extraordinary phenomena of hypnotism—which no doubt are in some degree related to the subject we are discussing, and which have been recognized for ages in the East—are forcing Western scientists to assume the existence of the so-called *secondary consciousness* in the body. The phenomena seem really inexplicable without the assumption of a secondary agency of some kind, and it every day becomes increasingly difficult *not* to use the word consciousness to describe it.

Let it be understood that I am not for a moment assuming that this secondary consciousness of the hypnotists is in all respects identical with the Cosmic Consciousness (or whatever we may call it) of the Eastern occultists. It may or may not be. The two kinds of consciousness may cover the same ground, or they may only overlap to a small extent. That is a question I do not propose to discuss. The point to which I wish to draw attention is that Western science is envisaging the possibility of the existence in man of another consciousness of some kind beside that with whose workings we are familiar. . . .

As a solid is related to its own surfaces, so, it would appear, is Cosmic Consciousness related to ordinary consciousness. The phases of the personal consciousness are but different facets of the other consciousness, and experiences which seem remote from each other in the individual are perhaps all equally near in the universal Space itself, as we know it,—may be practically annihilated in the consciousness of a larger space of which it is but the superficies; and a person living in London may not unlikely find that he has a back door opening quite simply and unceremoniously out in Bombay.[5]

Once described as "the Coleridge of psychical research," Frederic Myers (1843–1901) found evidence of a subliminal consciousness in telepathy, precognitive dreams, hypnosis, and the automatisms associated with trance mediumship and hysteria. Such phenomena were, for Myers, the strongest indications of the possibility of mental survival after death.

F. W. H. MYERS, "THE SUBLIMINAL CONSCIOUSNESS" (1892)

I suggest then, that the stream of consciousness in which we habitually live is not only consciousness which exists in connection with our organism. Our habitual or empirical consciousness may consist of a mere selection from a multitude of thoughts and sensations, of which some at least are equally conscious with those that we empirically know. I accord no primacy to my ordinary waking self, except that among my potential selves this one has shown itself the fittest to meet the needs of common life. I hold that it has established no further claim, and that it is perfectly possible that other thoughts, feelings, and memories, either isolated or in continuous connection, may now be actively conscious, as we say, 'within me,'—in some kind of co-ordination with my organism, and forming some part of my total individuality. I conceive it possible that at some future time, and under changed conditions, I may recollect all; I may assume these various personalities under one single consciousness, in which ultimate and complete consciousness the empirical consciousness which at this moment directs my hand may be only one element out of many.

Before we draw out the implications of such a statement, let us pause to consider the obvious reasons which a man may give for considering his empirical consciousness as identical with his total self.

The first remark of the ordinary reader will probably be that if there were in fact any other consciousness within him, he would certainly be aware of it.

This, however, is simply to beg the question. We must reply that the dicta of consciousness have already been shown to need correction in so many ways which the ordinary observer could never have anticipated,—the world of realities (so far as we can get at any intelligible notion of it) is so utterly unlike what our empirical consciousness suggests to us,—that we have no right to trust our consciousness, so to say, a step further than we can feel it;—to hold that anything whatever—even a separate consciousness in our own organism—can be proved *not* to exist, by the mere fact that we (as we know ourselves) are not aware of it.

But dropping this first untenable demurrer, a man may still give two reasons, which look valid enough, for supposing that there can be comparatively little psychical action going within him of which he cannot give an account. He may say, in the first place: 'The deeds which I have done in life—the movements which my body has made have been executed in obedience to the will of my conscious self. There has been no room for the operation of any imaginary will in the background'. And, in the second place, he may add: 'Besides this active life there has, of course, been a passive life as well. Besides the sensations and movements originating in my own frame, there have been sensations and movements impressed upon me from without. But all this, though I could not *control* it, I can nevertheless *remember*. I can feel sure that nothing of importance has happened to me which I cannot recall by voluntary act of recollection. Here again, therefore, there is no room for the operation of an imaginary memory beneath the threshold. In short, to put the matter in a nutshell, I receive my letters at my front door, and I give my orders in my library. Why should I suppose that my house is governed by an imaginary conspiracy in the kitchen!' . . .

For such is my hypothesis. I suggest that each of us is in reality an abiding psychical entity far more extensive than he knows—an individuality which can never express itself completely through any corporeal manifestation. The Self manifests through the organism; but there is always some part of the Self unmanifested; and always, as it seems, some power of organic expression in abeyance or in reserve. Neither can the player express all his thought upon the instrument, nor is the instrument so arranged that all its keys can be sounded at once. One melody after another may be played upon it; may,—as with the messages of duplex or of multiplex telegraphy,—simultaneously or with imperceptible intermissions, several melodies of instrumental capacity, as well as unexpressed treasures of informing thought.[6]

Swami Vivekananda (1863–1902) made his first visit to the West in 1893, as a speaker at the Parliament of World Religion in Chicago. His highly popular book Raja Yoga *resonated with fin-de-siècle harmonial religion, providing the foundations for modern yoga.*

SWAMI VIVEKANANDA, *RAJA YOGA* (1896)

In this country there are Mind-healers, Faith-healers, Spiritualists, Christian Scientists, Hypnotists, etc., and if we analyse these different groups we shall find that the background of each is this control of the *Prana*, whether they know it or not. If you boil all their theories down the residuum will be the same. It is the one and same force they are manipulating, only unknowingly. They have stumbled on the discovery of a force, and do not know its nature, but they are unconsciously using the same powers which the *Yogi* uses, and which come from *Prana*.

This *Prana* is the vital force in every being, and the finest and highest action of *Prana* is thought. This thought, again, as we see, is not all. There is also a sort of thought which we call instinct, or unconscious thought, the lowest plane of action. If a mosquito stings us, without thinking, our hand will strike it, automatically, instinctively. This is one expression of thought. All reflex actions of the body belong to this plane of thought. There is then a still higher plane of thought, the conscious. I reason, I judge, I think, I see the *pros* and *cons* of certain things; yet that is not all. We know that reason is limited. There is only a certain extent to which reason can go; beyond that it cannot reach. The circle within which it runs is very, very limited indeed. Yet, at the same time, we find facts rush into this circle. Like the coming of comets certain things are coming into this circle, and it is certain they come from outside the limit, although our reason cannot go beyond. The causes of the phenomena protruding themselves in this small limit are outside of this limit. The reason and the intellect cannot reach them, but, says the *Yogi*, that is not all. The mind can exist on a still higher plane, the super-conscious. When the mind has attained to the state, which is called *Samadhi*,—perfect concentration, super-consciousness—it goes beyond the limits of reason and comes face to face with facts which no instinct or reason can ever know. All these manipulations of subtle forces of the body, the different manifestations of *Prana*, if trained, give a push to the mind, and the mind goes up higher, and becomes super-conscious, and from that plane it acts.[7]

Freethinker and socialist Annie Besant (1847–1933) joined the Theosophical Society in 1889. A prolific lecturer and writer, her thoughts on the sub-conscious and super-conscious fused evolutionary theory and scientific psychology with spiritualism and Eastern religion.

ANNIE BESANT, *STUDY IN CONSCIOUSNESS* (1900)

The sympathetic system is a storehouse of traces left by long-past events—events not belonging to our present life at all, but events that passed hundreds of centuries ago, that occurred in long-past lives, when the Jivatma which is our Self was abiding in savage human bodies, and even in the bodies of animals. Many a causeless terror, many a midnight panic, many a surge of furious anger, many an impulse of vindictive cruelty, many a rush of passionate revenge, is flung up from the depths of that dark sea of the sub-conscious which rolls within us, concealing many a wreck, many a skeleton of our past. Handed down by the astral consciousness of the time to its physical instrument for putting into action, the ever-sensitive plate of the permanent atom has caught and photographed them, and has registered them in the recesses of the nervous system, life after life. The consciousness is off guard; or a strong vibration from another strikes us; or some event reproduces circumstances that start vibrations that arouse; in one way or another, the slumbering possibilities are awakened, and hurling itself upwards into the light of day comes the long-buried passion. There too hide the instincts which oft overpower reason, instincts that were once life-preserving efforts, or the results of experiences in which our body of the time perished, and the soul registered the result for future guidance. Instincts of love for the opposite sex, outcome of innumerable unions. Instincts of paternal and maternal love, poured out in many generations. Instincts of self-defence, developed in countless battles. Instincts of taking undue advantage, offspring of numberless cheatings and intrigues. And yet again there lurk there many vibrations that belong to events, and feelings, and desires, and thoughts of our present life, experienced and forgotten, but lying near the surface, ready for upcall. Time would fail to enumerate the contents of this relic-chamber of an immemorial past, containing old bones fit only for the dust-bin, side by side with interesting fragments of earlier days, with tools still useful for our present needs. Over the door of the relic-chamber is written: "Fragments of the Past". For the sub-consciousness belongs to the Past, as the waking-consciousness to the Present, as the super-consciousness to the Future.

Another part of the sub-conscious in us is composed of the contents of all the consciousnesses that use our bodies as fields of evolution—atoms, molecules, cells of many grades. Some of the queer spectres and dainty figures that arise from the sub-conscious in us do not belong to us at all, but are the dim gropings, and foolish fears, and pretty fancies, of the Units of consciousness at a lower stage of evolution than our own, that are our guests, inhabiting our body as a lodging-house.

In this part of the sub-conscious go on the wars, waged by one set of creatures in our blood against another set, which do not enter our consciousness, save when their results appear as diseases.[8]

Contra Freud, the novelist D. H. Lawrence (1885–1930) emphasized the vital and future-directed forces that were essential to the unconscious.

D. H. LAWRENCE, *PSYCHOANALYSIS AND THE UNCONSCIOUS* (1921)

It is necessary for us to know the unconscious, or we cannot live, just as it is necessary for us to know the sun. But we need not explain the unconscious, any more than we need explain the sun. We can't do either, anyway. We know the sun by beholding him and watching his motions and feeling his changing power. The same with the unconscious. We watch it in all its manifestations, its unfolding incarnations. We watch it in all its processes and its unaccountable evolutions, and these we register.

For though the unconscious is the creative element, and though, like the soul, it is beyond all law of cause and effect in its totality, yet in its processes of self-realization it follows the laws of cause and effect. The processes of cause and effect are indeed part of the working out of this incomprehensible self-realization of the individual unconscious. The great laws of the universe are no more than the fixed habits of the living unconscious.

What we must needs do is to try to trace still further the habits of the true unconscious, and by mental recognition of these habits break the limits which we have imposed on the movement of the unconscious. For the whole point about the true unconscious is that it is all the time moving forward, beyond the range of its own fixed laws or habits. It is no good trying to superimpose an ideal nature upon the unconscious. We have to try to recognize the true nature and then leave the unconscious itself to prompt new movement and new being—the creative progress.

What we are suffering from now is the restriction of the unconscious within certain ideal limits. The more we force the ideal the more we rupture the true movement. Once we can admit the known, but incomprehensible, presence of the integral unconscious; once we can trace it home in ourselves and follow its first revealed movements; once we know how it habitually unfolds itself; once we can scientifically determine its laws and processes in ourselves: then at last we can begin to live from the spontaneous initial prompting, instead of from the dead machine-principles of ideas and ideals. There is a whole science of the creative unconscious, the unconscious in its law-abiding activities. And of this science we do not even know the first term.[9]

In 1917, wealthy heiress and patron Mabel Dodge Luhan (1879–1962) married a peyote leader from Taos, New Mexico. Her second experience with peyote unfolded as a revelatory "expansion of consciousness," a coinage that would be embraced by a later generation of psychedelic explorers.

MABEL DODGE LUHAN, *INTIMATE MEMORIES* (1937)

The medicine ran through me, penetratingly. It acted like an organizing medium coordinating one part with another, so all the elements that were combined in me shifted like the particles in a kaleidoscope and fell into an orderly pattern. Beginning with the inmost central point in my own organism, the whole universe fell into place; I in the room and the room I was in, the old building containing the room, the cool wet night space where the building stood, and all the mountains standing out like sentries in their everlasting attitudes. So on and on into wider spaces farther than I could divine, where all the heavenly bodies were contented with the order of the plan, and system within system interlocked in grace. I was not separate or isolated anymore. The magical drink had revealed the irresistible delight of spiritual composition, the regulated relationship of one to all and all to one.

Was it this, I wondered, something like this, that artists are perpetually trying to find and project upon their canvases? Was this what musicians imagine and try to formulate? Significant Form!

I laughed there alone in the dark, remembering the favorite phrase that had seemed so hackneyed for a long time and that I had never really understood. Significant form, I whispered; why, that means that all things are really related to each other. These words had an enormous vitality and importance when I said them, more than they ever had afterwards when from time to time I approximately understood and realized their secret meaning after I relapsed into the usual dreamlike state of everyday life. . . .

And I learned that there is no single equilibrium anywhere in existence, and that the meaning and essence of balance is that it depends upon neighboring organisms, one leaning upon the other, one touching another, holding together, reinforcing the whole, creating form and defeating chaos.[10]

In the lexicon of Zen Buddhism, as popularized by the Buddhist monk and scholar D. T. Suzuki (1870–1966), mind expansion was also the keynote of satori.

D. T. SUZUKI, *INTRODUCTION TO ZEN BUDDHISM* (1934)

Without the attainment of *satori* no one can enter into the truth of Zen. *Satori* is the sudden flashing into consciousness of a new truth hitherto undreamed of. It is a sort of mental catastrophe taking place all at once, after much piling up of matters intellectual and demonstrative. The piling has reached a limit of stability and the whole edifice has come tumbling to the ground, when, behold, a new heaven is open to full survey. When the freezing point is reached, water suddenly turns into ice; the liquid has suddenly turned into a solid body and no more flows freely. *Satori* comes upon a man unawares, when he feels that he has exhausted his whole being. Religiously, it is a new birth; intellectually, it is the acquiring of a new viewpoint. The world now appears as if dressed in a new garment, which seems to cover up all the unsightliness of dualism, which is called delusion in Buddhist phraseology. . . .

Satori is the raison d'être of Zen without which Zen is no Zen. Therefore every contrivance, disciplinary or doctrinal, is directed toward *satori*. Zen masters could not remain patient for satori to come by itself; that is, to come sporadically or at its own pleasure. In their earnestness to aid their disciples in the search after the truth of Zen their manifestly enigmatical presentations were designed to create in their disciples a state of mind which would more systematically open the way to enlightenment.[11]

For Carl Jung (1875–1961), the pursuit of self-knowledge lay in understanding the primordial archetypes that were the ego's puppet masters.

CARL GUSTAV JUNG, *AION* (1951)

Whereas the contents of the personal unconscious are acquired during the individual's lifetime, the contents of the collective unconscious are invariably archetypes that were present from the beginning. . . . The

archetypes most clearly characterized from the empirical point of view
are those which have the most frequent and the most disturbing influence on the ego. These are the *shadow*, the *anima*, and the *animus*. The
most accessible of these, and the easiest to experience, is the shadow, for
its nature can in large measure be inferred from the contents of the personal unconscious. The only exceptions to this rule are those rather rare
cases where the positive qualities of the personality are repressed, and the
ego in consequence plays an essentially negative or unfavourable role.

The shadow is a moral problem that challenges the whole ego-
personality, for no one can become conscious of the shadow without considerable moral effort. To become conscious of it involves recognizing the
dark aspects of the personality as present and real. This act is the essential
condition for any kind of self-knowledge, and it therefore, as a rule, meets
with considerable resistance. Indeed, self-knowledge as a psychotherapeutic measure frequently requires much painstaking work extending
over a long period. Closer examination of the dark characteristics—that
is, the inferiorities constituting the shadow—reveals that they have an
emotional nature, a kind of autonomy, and accordingly an obsessive or,
better, possessive quality. Emotion, incidentally, is not an activity of the
individual but something that happens to him. Affects occur usually
where adaptation is weakest, and at the same time they reveal the reason
for its weakness, namely a certain degree of inferiority and the existence
of a lower level of personality. On this lower level with its uncontrolled
or scarcely controlled emotions one behaves more or less like a primitive,
who is not only the passive victim of his affects but also singularly incapable of moral judgement.[12]

*Parapsychological researcher J. B. Rhine (1895–1980) proposed that psi
phenomena (telepathy, clairvoyance, and precognition) belonged to an
operationally closed part of the human personality.*

JOSEPH BANKS RHINE, *NEW WORLD OF THE MIND* (1953)

Psi, then, is normal. That conclusion helps to place it in the general psychological framework. But the most significant and revealing characteristic of psi is the fact that its operation is entirely unconscious. . . . It is
not merely one of those transiently subconscious processes that go on

practically all the time in everyone, mental activities that can be consciously recovered if one tries and knows how to do so. It is not even one of those more hidden kinds which the psychiatrist can bring to the surface if, because of an unhealthy attitude, control over them has been lost. Neither is the psi experience similar to the dissociated or submerged section of consciousness which can be allowed to operate itself during sleep, as in dreams; rather, psi is simply not capable of being dragged into consciousness unconverted and direct. Such seems to be the case as it stands today. There are not even any good leads to conscious control in sight.[13]

Armenian mystagogue George Ivanovitch Gurdjieff (1866–1949) instructed his pupil P. D Ouspensky (1878–1947) on the awakening of higher states of consciousness through self-observation.

P. D. OUSPENSKY, *THE FOURTH WAY* (1957)

[S]elf-observation is necessary before we really recognize the fact that we are not conscious; that we are conscious only potentially. If we are asked, we say, 'Yes, I am', and for that moment we are, but the next moment we cease to remember and are not conscious. . . .

So, at the same time as self-observing, we try to be aware of ourselves by holding the sensation of 'I am here'—nothing more. And this is the fact that all Western psychology, without the smallest exception, has missed. Although many people came very near to it, they did not recognize the importance of this fact and did not realize that the state of man as he is can be changed—that man can remember himself, if he tries for a long time.

It is not a question of a day or a month. It is a very long study, and a study of how to remove obstacles, because we do not remember ourselves, we are not conscious of ourselves, owing to many wrong functions in our machine, and all these functions have to be corrected and put right. When most of these functions are put right, these periods of self-remembering will become longer and longer, and if they become sufficiently long, we shall acquire two new functions. With self-consciousness, which is the third state of consciousness, we acquire a function which is called *higher emotional*, although it is equally intellectual, because on this level there is no difference between intellectual and emotional such as there is on the ordinary level. And when we come to the state of objective consciousness we acquire another function which is called *higher mental*. Phenomena of what I call supernormal psychology belong to these two functions.[14]

Psychedelics provided Aldous Huxley (1894–1963) with a conduit to the mystical and transfigured realms of Catholicism's Beatific Vision, the Vedic Sat Chin Ananda, and the Voidness of Tibetan Buddhism. After his first dose of mescaline, Huxley concluded that the function of the brain and nervous system was in ordinary circumstances, like language itself, largely eliminative.

ALDOUS HUXLEY, *THE DOORS OF PERCEPTION* (1954)

To make biological survival possible, Mind at Large has to be funneled through the reducing valve of the brain and nervous system. What comes out at the other end is a measly trickle of the kind of consciousness which will help us to stay alive on the surface of this Particular planet. To formulate and express the contents of this reduced awareness, man has invented and endlessly elaborated those symbol-systems and implicit philosophies which we call languages. Every individual is at once the beneficiary and the victim of the linguistic tradition into which he has been born—the beneficiary inasmuch as language gives access to the accumulated records of other people's experience, the victim in so far as it confirms him in the belief that reduced awareness is the only awareness and as it bedevils his sense of reality, so that he is all too apt to take his concepts for data, his words for actual things. That which, in the language of religion, is called "this world" is the universe of reduced awareness, expressed, and, as it were, petrified by language. The various "other worlds," with which human beings erratically make contact are so many elements in the totality of the awareness belonging to Mind at Large. Most people, most of the time, know only what comes through the reducing valve and is consecrated as genuinely real by the local language. Certain persons, however, seem to be born with a kind of by-pass that circumvents the reducing valve. In others temporary by-passes may be acquired either spontaneously, or as the result of deliberate "spiritual exercises," or through hypnosis, or by means of drugs.[15]

*Teilhard de Chardin's (1881–1955) heady ferment of evolutionary science
and Christian mysticism introduced the concept of the noosphere, an emergent
sphere of thought that the French Jesuit and paleontologist believed would be
humanity's post-biological home.*

TEILHARD DE CHARDIN, *THE PHENOMENON OF MAN* (1955)

The universe is a collector and conservator, not of mechanical energy, as
we supposed, but of persons. All round us, one by one, like a continual
exhalation, 'souls' break away, carrying upwards their incommunicable
load of consciousness. One by one, yet not in isolation. Since, for each
of them, by the very nature of Omega, there can only be one possible
point of definitive emersion—that point at which, under the synthesising
action of personalising union, the noosphere (furling its elements upon
themselves as it too furls upon itself) will reach collectively its point of
convergence—at the 'end of the world.' . . .

Now when sufficient elements have sufficiently agglomerated, this
essentially convergent movement will attain such intensity and such
quality that mankind, taken as a whole, will be obliged—as happened to
the individual forces of instinct—to reflect upon itself at a single point;
that is to say, in this case, to abandon its organo-planetary foothold so as
to shift its centre on to the transcendent centre of its increasing concentration. This will be the end and the fulfillment of the spirit of the earth.

The end of the world: the wholesale internal introversion upon itself
of the noosphere, which has simultaneously reached the uttermost limit
of its complexity and its centrality.

The end of the world: the overthrow of equilibrium, detaching the
mind, fumed at last, from its material matrix, so that it will henceforth
rest with all its weight on God-Omega.[16]

*The religious historian Mircea Eliade (1907–1986) navigated the spiritual
technologies of yoga for a new generation of seekers.*

MIRCEA ELIADE, *YOGA: IMMORTALITY AND FREEDOM* (1958)

THE point of departure of Yoga meditation is concentration on a single
object; whether this is a physical object (the space between the eyebrows,

the tip of the nose, something luminous, etc.), or a thought (a meta-physical truth), or God (Ishvara) makes no difference. This determined and continuous concentration, called *ekāgratā* ("on a single point"), is obtained by integrating the psychomental flux (*sarvārthatā*, "variously directed, discontinuous, diffused attention"). This is precisely the definition of yogic technique: *yogah cittavrtti-nirodhyah*, i.e., the yoga is the suppression of psychomental states (*Yoga-sutrās*, 1, 2).

The immediate result of *ekāgratā*, concentration on a single point, is prompt and lucid censorship of all the distractions and automatisms that dominate—or, properly speaking, compose—profane conscious-ness. Completely at the mercy of associations (themselves produced by sensations and the *vāsanās*), man passes his days allowing himself to be swept hither and thither by an infinity of disparate moments that are, as it were, external to himself. The senses or the subconscious continu-ally introduce into consciousness objects that dominate and change it, according to their form and intensity. Associations disperse conscious-ness, passions do it violence, the "thirst for life" betrays it by projecting it. Even in his intellectual efforts, man is passive, for the fate of secular thoughts (controlled not by *ekāgratā* but only by fluctuating moments of concentration, *kshiptavikshiptas*) is to be thought by objects. Under the appearance of thought, there is really an indefinite and disordered flickering, fed by sensations, words, and memory. The first duty of the yogin is to think—that is, not to let himself think. This is why Yoga prac-tice begins with *ekāgratā*, which damns the mental stream. . . . Through *ekāgratā* one gains a genuine will—that is, the power freely to regulate an important sector of biomental activity.[17]

Echoing the evolutionary spirituality of Chardin, the self-styled "philosophical entertainer" Alan Watts (1915–1973) called time on the Freudian unconscious.

ALAN WATTS, *THE BOOK: ON THE TABOO AGAINST KNOWING WHO YOU ARE* (1966)

The most strongly enforced of all known taboos is the taboo against knowing who or what you really are behind the mask of your appar-ently separate, independent, and isolated ego. I am not thinking of Freud's barbarous Id or Unconscious as the actual reality behind the façade of personality. Freud, as we shall see, was under the influence of a nineteenth-century fashion called "reductionism," a curious need to put down human culture and intelligence by calling it a fluky by-product of blind and irrational forces. They worked very hard, then, to prove that grapes can grow on thornbushes.

As is so often the way, what we have suppressed and overlooked is something startlingly obvious. The difficulty is that it is so obvious and basic that one can hardly find the words for it. The Germans call it a *Hintergedanke*, an apprehension lying tacitly in the back of our minds which we cannot easily admit, even to ourselves. The sensation of "I" as a lonely and isolated center of being is so powerful and commonsensical, and so fundamental to our modes of speech and thought, to our laws and social institutions, that we cannot experience selfhood except as something superficial in the scheme of the universe. I seem to be a brief light that flashes but once in all the aeons of time—a rare, complicated, and all-too-delicate organism on the fringe of biological evolution, where the wave of life bursts into individual, sparkling, and multicolored drops that gleam for a moment only to vanish forever. Under such conditioning it seems impossible and even absurd to realize that myself does not reside in the drop alone, but in the whole surge of energy which ranges from the galaxies to the nuclear fields in my body.[18]

California's Esalen Institute, the center of the Human Potential Movement, was home to the German-born psychotherapist Fritz Perls (1893–1970) in the late 1960s. Perls' approach to mindfulness—aided by a litany of Gestalt prayers—became the guiding credo of experiential groups and communities across the country.

FRITZ PERLS, *THE GESTALT APPROACH* (1973)

Many of the neurotic's difficulties are related to his unawareness, his blind spots, to the things and relationships he simply does not sense. And therefore, rather than talking of the unconscious, we prefer to talk about the at-this-moment-unaware. This term is much broader and wider than the term "unconscious". This unawareness contains not only repressed material, but material which never came into awareness, and material which has faded or has been assimilated or has been built into larger gestalts. The unaware includes skills, patterns of behaviour, motoric and verbal habits, blind spots, etc. . . .

Gestalt therapy is an experiential therapy, rather than a verbal or an interpretive therapy. We ask our patients not to talk about their traumas and their problems in the removed area of the past tense and memory, but to re-experience their problems and their traumas—which are their unfinished situations in the present—in the here and now. If the patient is finally to close the book on his past problems, he must close it in the present. For he must realize that if his past problems were really past, they would no longer be problems—and they certainly would not be present. In addition, as an experiential therapy, the Gestalt technique demands of the patient that he experience as much of himself as he can, that he experience himself as fully as he can in the here and now. We ask the patient to become aware of his gestures, of his breathing, of his emotions, of his voice, and of his facial expressions as much as of his pressing thoughts. We know that the more he becomes aware of himself, the more he will learn about what his self is.[19]

Theodore Roszak (1933–2011) on the shaman as culture hero and explorer of superconsciousness.

THEODORE ROSZAK, *THE MAKING OF A COUNTER CULTURE* (1969)

In his inspired taletelling we might find the beginnings of mythology, and so of literature; in his masked and panted impersonations, the origin of the drama; in his entranced gyrations, the first gestures of the dance. He was—besides being artist, poet, dramatist, dancer—his people's healer, moral counsellor, diviner, and cosmologer. . . .

The techniques by which shamans undertake their psychic adventures are many; they may make use of narcotic substances, dizziness, starvation, smoke inhalation, suffocation, hypnotic drum and dance rhythms, or even the holding of one's breath. One recognizes at once in this trance-inducing repertory a number of practices which underlie the many mystical traditions of the world: the practices of oracles, dervishes, yogis, sibyls, prophets, druids, etc.—the whole heritage of mystagoguery toward which the beat-hip wing of our counter culture now gravitates.

By such techniques, the shaman cultivates his rapport with the non-intellective sources of the personality as assiduously as any scientist trains himself to objectivity, a mode of consciousness at the polar extreme from that of the shaman. Thus the shaman is able to diffuse his sensibilities through his environment, assimilating himself to the surrounding universe. He enters wholly into the grand symbiotic system of nature, letting its currents and nuances flow through him. He may become a keener student of his environment than any scientist. He may be able to taste rain or plague on the wind. He may be able to sense the way the wild herds will move next or how the planting will go in the season to come.

The shaman, then, is one who knows that there is more to be seen of reality than the waking eye sees. Besides our eyes of flesh, there are eyes of fire that burn through the ordinariness of the world and perceive the wonders and terrors beyond.[20]

Biofeedback offers a technological short cut to the "alpha state."

GAY LUCE AND ERIK PEPER, "MIND OVER BODY, MIND OVER MIND" (1971)

[M]any apparently healthy young students . . . in need of serenity and in search of salvation have trooped to the laboratories that do research in "EEG feedback". The EEG machine, an electroencephalograph, is an amplifier system that receives very small shifts in electric potential that occur within the brain and are detected by electrodes on the surface of the head. The changing patterns of the brain waves are not understood, but seem to represent continuous shifts within brain cells from positive to negative charge. The bioelectric shifts, mightily amplified, drive a row of inked pens on old-fashioned laboratory machines. Each pen, moved by a signal from a part of the brain, swings up when the charge is negative and down when it is positive. As a continuous sheet of graph paper slides beneath the oscillating pens, the up-and-down pen movements are traced out as brain waves. Thus brain waves are an artifact of the way we record brain activity—a convenience, not of nature.

For convenience, the common brain-wave patterns observed in normal people during waking and sleep have been divided into categories,
according to the speed of change (frequency) and the amount of voltage in the change (amplitude). When a person is awake, the voltage is
very low, and waves are typically irregular and changing fast—high frequency. However, as a person relaxes, perhaps unfocusing his eyes and
drifting for a moment, the irregular scrawl on the EEG paper will show
a burst of regular waves somewhere between 8 and 13 cycles per second,
of high voltage (maybe 60 microvolts). This rhythm, which is observed
occasionally in sleep, and during yoga and Zen meditation, is known as
the alpha rhythm. Because it has been recorded in people while they were
experiencing pleasurable relaxation or meditation states—and because
industry is exploiting the fad—a cult has been growing up around the
"alpha state."[21]

*Transpersonal psychology looked to the near-death experience as confirmation
of consciousness outside the body.*

MICHAEL GROSSO, *THE MYTH OF THE NEAR-DEATH JOURNEY* (1991)

The modern rational mind sees death as the end, a "grim reaper," a
cul-de-sac; but . . . typical near-death experience[s]—emerging spontaneously as they seem to from a deep, perhaps collective layer of mind—
tell a different story: death, seen through the eyes of the unconscious,
becomes a boat, a voyage, a journey into the light, a love embrace with
the universe. . . .

So this is one way to think of the meaning of the near-death experience: in terms of what it is doing to us, reviving—not in a rational or
objective way, but at a grass roots level of popular consciousness—the
living world of the mythic journey. The new (yet very old) myth that
is crystallizing out of modern NDEs—millions if the polls are right—is
telling us in no uncertain terms that death is a journey. The image of
death emerging from millions of unconscious minds is clear: it radically
reverses "logic," common sense, and ordinary science. It substitutes light
for dark, joy for grief, movement for stagnation; it affirms the claims of

ancient visionaries from Zoroaster to Saint Paul and Plotinus. And it is doing so in the form of a living myth. . . .

Millions of people having the same experience are telling us to trust ourselves—not our ideas or our beliefs, but the source of life that lies coiled in our own inner depths. There is a tremendous life force, a life light, a life tide within. The NDE is telling us to trust that light, that the way has been tried, that where things look gloomiest, darkest, most hopeless, there is really a hidden spring of light ahead. The NDE is calling us to "follow our bliss," to use the now famous words of Joseph Campbell.[22]

The notion of akasha was a watchword for early Theosophists who proposed the existence of a cosmic thought-field containing "all the desires and earth experiences of our planet." For the Hungarian-born philosopher of science Ervin Laszlo (1932–), the Akashic record, or A-field, remains key to unlocking all shades of scientific mystery, from quantum entanglement to near-death experiences and the non-random nature of evolution.

ERVIN LASZLO, *SCIENCE AND THE AKASHIC FIELD* (2004)

A cosmic field that underlies and links all things in the world is a perennial intuition, present in traditional cosmologies and metaphysics. The ancients knew that space is not empty: it is the origin and the memory of all things that exist and have ever existed. But this knowledge was based on philosophical or mystical insight, the fruit of private and unrepeatable if often seemingly indubitable experience. The current rediscovery of the Akashic Field reinforces qualitative human experience with quantitative data generated by science's experimental method. The combination of unique personal insight and interpersonally observable and repeatable experience gives us the best assurance we can have that we are on the right track: that a cosmic information field connects organisms and minds in the biosphere, and particles, stars, and galaxies throughout the cosmos.

Nature's information field is now being rediscovered at the cutting edge of the sciences. It emerges as a powerful fable and—as sustained research deepens and specifies the theory of the A-field—as a main pillar of the scientific world picture of the twenty-first century. This will profoundly change our concept of ourselves and of the world.

The rediscovery of the A-field will also change our world itself. When people realize that the age-old intuition that space does not separate things but links them has a bona fide scientific explanation, the genius for innovation inherent in modern civilization will find ways to make practical use of it. As people learn to work with the A-field, untold ways will come to light for beaming active and effective "in-formation" from one place to another, instantly and without the expenditure of energy. This will not only enable quantum computation, but also pave the way to an entire series of technological breakthroughs.[23]

EPILOGUE

Sigmund Freud did not believe in accidents. A wrongly dated letter. A mis-remembered name. A bungled handshake. Each of these apparently inno-cent mistakes was, Freud wrote in *The Psychopathology of Everyday Life*, a tell-tale act capable of making the "unknown known to consciousness."

First published in book form in 1904, *The Psychopathology of Every-day Life* was my introduction to the vagaries of the unconscious. Prior to finding a copy in our school library, aged thirteen or fourteen, my only contact with Freud had been at my local barbershop—a cartoonish poster of the bearded and spectacled sage which, on second glance, fea-tured an outstretched, naked woman. This slightly salacious introduction may have helped draw me to the slim paperback, but its content proved as sketchy as his cartoon profile. Was it really possible to trace every-day blunders, verbal or physical, to unconscious fears and frustrations? Were the tongue-tied and accident-prone routinely ambushed by unreal-ized thoughts? And what had happened at Berggasse 19, Freud's Vienna home, to make him think that when servants broke his most valuable items they were acting out of "foolish hostility" towards them.

Two decades on, when I began to take a more serious interest in psy-choanalysis and the history of the unconscious, a new generation of scholars was following Henri Ellenberger's lead in examining the intel-lectual roots of dynamic psychiatry and depth psychology. Freud's char-acter and reputation were on the line, as his case histories unraveled to reveal a catalogue of lies and obfuscation, all compounded by a string of therapeutic failures. On close inspection, the psychanalytic movement began to resemble a kind of psychotherapeutic Ponzi scheme, and its first tranche of medical followers readily played the part of over-invested franchisees, fearful that criticism or poor publicity might come to upset the delicate economy of the fifty-minute hour.

The charge sheet did not end here. To burnish his exclusive claim to have cracked the Enigma machine of the unconscious—decoding the ciphertext of neuroses, dreams and commonplace blunders—there was little doubt that Freud had strategically disavowed knowledge of his most important philosophical forebears, particularly Schopenhauer. (Imagine a political economist who openly claimed to have avoided reading Marx, or an evolutionary biologist deciding that Darwin might prove a distraction). Time and again, rival theorists and co-workers had been looked over. In the *Psychopathology of Everyday Life*, Freud barely acknowledged the possibility that motors accidents and verbal slips might be the outcome of psychological mechanisms that were already clinically and experimentally well attested. Divided attention, attentional drift, the power of association, the mind's runaway tendency to operate automatically in familiar situations—surely, all these processes offered better explanations for most mistakes of speech and action than Freud's arcane flights of interpretation.

Debate and controversy, neglect and re-discovery, have always closely followed the intellectual history of the unconscious. Take, for example, Plato's thoughts on the native capacity of the mind, sometimes known as a theory of innateness. Sidelined by the Aristotelian system that came to dominate Western philosophy for well over two millennia, nativism returned to the fore in the Enlightenment, when it was revived by Descartes and then fiercely rebutted by Locke and Leibniz. Stalled circles of this kind can be seen throughout the social and intellectual history of the unconscious, with theories and phenomena disappearing and then returning into view.

Over the course of the nineteenth and twentieth centuries, all schools of the unconscious were riven by conflict and dispute. The concept of "unconscious cerebration" met with vigorous criticism from philosophers who were ill-disposed to the notion of thought without awareness, and from theologians who bristled at its assault on human agency and free will. Hypnosis, rejected by Freud in favor of free association, became a bona fide tool for the treatment of trauma and psychological disorders, but there was no academic consensus on the mental status of the trance

state. The list goes on, taking in the brouhaha provoked by behaviorism, the Oedipus complex, Jung's collective unconscious, new laboratory
theories of sleep and dreams, subliminal priming, recovered memories,
and unconscious bias.

Similar uncertainty applies to research and theorizing on the unconscious processes from other disciplines. Brain imaging techniques such
as PET scans and fMRIs have allowed neuroscientists to repeatedly dem-
onstrate that experimental subjects may seemingly experience emotions
and perceptions of which they have no conscious awareness, but grainy
snapshots of brain regions that participate in mediating such processes
do not deliver any direct understanding of their psychological nature. A
hundred and fifty years after the birth of psychoanalysis and scientific
psychology, the warning that William James delivered back in 1890, that
the great danger for psychology was that it would become too materialis-
tic, remains relevant for research on unconscious mental activity.

As for today's popularizers of the unconscious or subconscious, too
much self-help literature has a distinctly jaded quality. From golden rules
for habit formation to daily affirmations that promise to reprogram the
subconscious mind, most of what is now passed off as "science" by its
print and clickbait promoters has little or no evidential foundation. Magical thinking of this kind takes us back to the 1920s, when faith in the
power of the subconscious as a source of self-improvement widely seeded
the throwaway promise of health and success.

We can already discern some of the dramas and insights the future
may hold for the unconscious mind. Neuro-inspired AI models are opening new horizons for studies of implicit learning, creativity and behavior modification. Cognitive science is delivering more detailed maps to
better understand how the burden of processing of sensory information
takes place outside the limited bandwidth of consciousness, and how
decision-making relies on an inventory of offline mental shortcuts. And
findings from the psychological laboratory provide strong intimations as
to how our emotions are constantly primed by the secret alchemy of rapport, contagion and environmental cues.

Yet the reality and complexity of our unconscious lives remains as
elusive as ever. Well over a century has passed since Freud, Janet, and
other clinicians began to untangle the psychological roots of hysteria,

and our current understanding of somatization and psychogenic illness—the ways in which trauma, stress, and life problems are transformed into legitimate physical symptoms such as pain and fatigue—remains limited. To confront the forces that shape and underwrite our present-day idioms of distress requires more than an understanding of their underlying biology and psychology. Such an endeavor requires a medical sociology of the unconscious—a discipline that would seek to understand how, in the words of medical historian Anne Harrington, "human beings seem to be invested with a developed capacity to mold their bodily experiences to the norms of their cultures . . . [to] learn the scripts about what kind of things should be happening to them as they fall ill."[1] A deeper understanding of this shifting semaphore of psychogenic illness is one way in which we can begin to grasp the unconscious coupling implied by the dash between mind-body.

Whatever we go on to learn experimentally, clinically, or computationally about the unconscious mind, the blackboard of history carries one lasting lesson: the mysteries of the unconscious will always be renewed, its puzzles often becoming deeper. And on those rare occasions that our mental blindfolds do slip—affording a cryptic glimpse of a shadow thought, memory or emotion—we may simply never know whether this was by accident or design.

ACKNOWLEDGMENTS

For permission to reprint copyright material from the following the publishers gratefully acknowledge the following:

Noam Chomsky for an extract from *Reflections on Language* (Pantheon, 1975); Peter Connor and Phil Baker for an extract from *Dreams and How to Guide Them* by Leon Hervey de Saint-Denys (Strange Attractor Press, 2025); Dover Publications for an extract from *Fads and Fallacies in the Name of Science* by Martin Gardner (Dover Publications, 1957); Howard Gardner for an extract from *The Shattered Mind* (Routledge & Kegan Paul, 1977); Harvard University Press for an extract from *Children's Dreaming and The Development of Consciousness* by David Foulkes (Harvard University Press, 1999); Princeton University Press for an extract from *The Collected Works of C. G. Jung, Vol. 9* (Princeton University Press, 1981); W. W. Norton for an extract from *Soul Machine: The Invention of the Modern Mind* by George Makari (W. W. Norton, 2015); W. W. Norton for an extract from *The Living Brain* by William Grey Walter, *The Living Brain* (W. W. Norton, 1953).

NOTES

PROLOGUE

1. William James, *The Varieties of Religious Experience: A Study in Human Nature* (New York: Penguin, 1982), 483–484.

2. René Descartes, "Meditations on First Philosophy," in *The Philosophical Writings of Descartes*, vol. 1, ed. John Cottingham, Robert Stoothoff, and Dugald Murdoch (Cambridge: Cambridge University Press, 1985), 171.

3. G. W. Leibniz, *New Essays on Human Understanding*, ed. Peter Remnant and Jonathan Bennett (Cambridge: Cambridge University Press, [1765], 1996), 53–54.

4. Friedrich Nietzsche, *Beyond Good and Evil*, trans. Walter Kaufmann (Vintage, New York Books, [1886], 1966), 68.

5. Friedrich Nietzsche, *The Gay Science*, trans. Josefine Nauckhoff (Cambridge: Cambridge University Press [1882] 2007), 213–214.

6. F. W. J. Schelling, *System of Transcendental Idealism*, trans. Peter Heath (Charlottesville: University Press of Viginia [1800] 1978), 209.

7. Gustave Le Bon, *The Crowd: A Study of the Popular Mind* (New York: Dover Publications, 2002), 6.

8. William James, *The Principles of Psychology*, 1 (New York: Henry Holt, [1890] 1918), 163.

9. Edward Scripture, *Thinking, Feeling, Doing* (Meadville PA; Flood and Vincent, 1895), 62.

10. William James, "The Confidences of a Psychical Researcher," in *The Writings of William James*, ed. J. McDermot (Chicago; The University of Chicago Press [1909] 1977), 787–799.

11. Sigmund Freud, *The Unconscious*, trans. Graham Frankland (London; Penguin, [1915] 2005), 49.

12. J. A. Bargh, "The Automaticity of Everyday Life," in *The Automaticity of Everyday Life: Advances in Social Cognition*, vol. 10, ed. R. S. Wyer Jr. (Mahwah, NJ: Lawrence Erlbaum Associates, 1997), 1–61.

13. Kenneth Burke, *Language as Symbolic Action: Essays on Life and Letters* (Berkeley: University of California Press, 1966), 45.

CHAPTER 1

1. Heraclitus, quoted in John Burnet, *Early Greek Philosophy* (London: Adam and Charles Black, 1908), 153.

2. Aristides, quoted in Norman Mackenzie, *Dreams and Dreaming* (London: Aldus Books, 1965), 45.

3. Paracelsus, quoted in Norman Mackenzie, *Dreams and Dreaming*, 72.

4. J. Allan Hobson, *Dream Life: An Experimental Memoir* (Cambridge, MA: MIT Press, 2011), 151.

5. Aristotle, "On Parts of Animals," in *The Complete Works of Aristotle*, ed. Jonathan Barnes (Princeton: Princeton University Press, [350 BCE] 1995), 994.

6. Hippocrates, *Hippocrates, Volume IV*, trans. W. H. S. Jones (Cambridge, MA: Harvard University Press ([400 BC] 1931), 423–427.

7. Artemidorus, *Artemidorus' Oneirocritica: Text, Translation, and Commentary*, trans. Daniel Harris-McCoy (Oxford: Oxford University Press, [2nd Century AD] 2012), 84–85.

8. Synesius of Cyrene, *The Essays and Hymns of Synesius of Cyrene* (Vol. 2), trans. Augustine Fitzgerald (London: Oxford University Press, [405 AD] 1930), 354–355.

9. St. Augustine, *Confessions*, trans. Henry Chadwick (Oxford: Oxford University Press, [397–400AD] 1998), 203.

10. Thomas Hobbes, *Leviathan*, ed. J. C. A. Gaskin (Oxford: Oxford University Press, ([1651] 1998), 12.

11. Erasmus Darwin, *Zoonomia, Vol. 1* (Boston: Thomas and Andrews, 1803), 158–166.

12. Roger A. Ekirch, *At Day's Close: A History of Nighttime* (New York: Norton, 2005), 334–335.

13. Charles Mackay, *Extraordinary Popular Delusions and The Madness of Crowds* (London: Wordsworth Books, [1841] 1995), 295.

14. Leon Hervey de Saint-Denys, *Dreams and How to Guide Them*, trans. Peter Connor, ed. Phil Baker (London: Strange Attractor Press, 2023), 327–328.

15. Edward B. Tylor, *Primitive Culture: Research into the Development of Mythology, Philosophy, Religion, Art and Custom, Vol. 1* (London: John Murray, 1871), 397–400.

16. Frances Power Cobbe, "Dreams as Illustration of Involuntary Cerebration," in *Darwinism in Morals and Other Essays* (London: William and Norgate, 1872), 338–339.

17. Friedrich Nietzsche, *Human, All Too Human*, trans. R. J. Hollingdale (Cambridge: Cambridge University Press ([1878] 1996), 16–17.

18. William Dean Howells, "True, I talk of Dreams," in *Oxford Book of American Essays*, ed. Brander Matthews (New York: Oxford University Press [1895] 1914), 308.

19. James Sully, "The dream as a revelation," *Fortnightly Review* no. 59 (1893): 354–365.

20. Sigmund Freud, *The Interpretation of Dreams: The Complete and Definitive Text*, ed. and trans. James Strachey (Basic Books: New York [1900] 2010), 518–519 and 562–563.

21. Havelock Ellis, *The World of Dreams* (Boston: Houghton Mifflin Company [1911] 1922), 278–279.

22. Henri Bergson, *Dreams*, trans. Edwin E Slosson (New York: B. W. Huebsch, 1914), 50.

23. Mary Arnold-Forster, *Studies in Dreams* (London: George Allen and Unwin, 1921) 57–58.

24. Calvin S. Hall, "What People Dream About," *Scientific American* 184, no. 5 (May 1951), 60–63.

25. William C. Dement, *Some Must Watch While Some Must Sleep* (New York: Norton, 1974), 24–25.

26. J. Allan Hobson, *Dream Life: An Experimental Memoir* (Cambridge, MA: MIT Press, 2011), 151–152.

27. Francis Crick and Graeme Mitchison, "The Function of Dream Sleep," *Nature* 304, no. 5922, (July 1983): 111–114.

28. Michel Jouvet, *The Paradox of Sleep* (Cambridge, MA: MIT Press, 1999), 140–142.

29. David Foulkes, *Children's Dreaming and The Development of Consciousness* (Cambridge, MA: Harvard University Press 1999), 116–117.

30. Darian Leader, *Why Can't We Sleep* (London: Hamish Hamilton, 2019), 96–98.

CHAPTER 2

1. Henry Holland, *Medical Notes and Reflections* (London: Longman, 1839), 351.

2. Angelo Mosso, *Fear*, trans. E. Lough and F. Kiesow (London: Longman, Green, [1891] 1895), 78.

3. Aristotle, "On Sleep," in *The Complete Works of Aristotle*, ed. Jonathan Barnes (Princeton, NJ: Princeton University Press, [350 BCE] 1995), 721–728.

4. Guy Claxton, *The Wayward Mind* (London: Abacus, 2002), 221.

5. Virginia Woolf, "Modern Fiction," in *The Common Reader* ([1919], 1925; reprint, Kolkata: Maulana Azad College), 4.

6. Samuel Johnson, "On Sleep," *The Idler*, no. 39 (1753).

7. Michel Montaigne, *The Complete Works of Montaigne*, trans. D. M. Frame (Stanford: Stanford University Press ([1580 BC] 1957), 668.

8. Miguel de Cervantes, *Don Quixote De La Mancha*, trans. Charles Jarvis (Oxford: Oxford University Press, [1605] 2008), 313–314.

9. Ralph Cudworth, *A Treatise Concerning Eternal and Immutable Morality*, ed. Sarah Hutton (Cambridge: Cambridge University Press, [1731] 1996), 67–72.

10. Jean-Jacques Rousseau, *Reveries of the Solitary Walker*, trans. Russell Gouldbourne (Oxford: Oxford University Press, ([1782] 2011), 55–56.

11. William Hazlitt, "On Dreams," in *The Plain Speaker, Volume 1* (London: Henry Colburn, 1826) 36–39.

12. William Godwin, *Thoughts on Man* (London: Effingham Wilson, 1831), 149–153.

13. Jacques-Joseph Moreau, *Hashish and Mental Illness*, trans. Gordon J. Barnett (New York: Raven Press, [1845] 1975), 33–34.

14. Pierre Janet, *The Mental State of Hystericals*, ed. Caroline Rollin Corson (New York: G. P. Putnam, [1892] 1901), 201–202.

15. Marie de Manaciene, *Sleep: Its Physiology, Pathology, Hygiene, and Psychology* (London: Walter Scott Ltd, 1897), 202–203.

16. Theodate L. Smith, "The Psychology Of Day Dreams," *The American Journal of Psychology* 15, no. 4, (October 1904): 469–474.

17. Herbert Silberer, "Report on a Method of Eliciting and Observing Certain Symbolic-Hallucination Phenomena," in *Organization and Pathology of Thought*, ed. and trans. David Rapaport (New York: Columbia University Press, [1909] 1951), 195–198.

18. André Breton, *What Is Surrealism?* ed. Franklin Rosemont, trans. David Gascoyne (New York: Monad Press, [1934] 1978) 120–121.

19. E. M. Cioran, *On the Heights of Despair*, trans. Ilinca Zarifopol-Johnston (Chicago: University of Chicago Press, [1934] 1992), 85.

20. Vladimir Nabokov, *Speak, Memory* (New York: Vintage Books [1947] 1989), 33–34.

21. James Thurber, *My World—and Welcome to It* (New York: Harcourt, Brace and Company ([1939] 1942), 72–73.

22. William Grey Walter, *The Living Brain* (Harmondsworth: Penguin, 1953), 202–203.

23. Louis J. West. et al., "The Psychosis of Sleep Deprivation," *Annals of the New York Academy of Sciences* 96, no. 5 (1962): 68–69.

24. Ernst Bloch, *The Principle of Hope, Vol. 1*, trans. Neville Plaice, Stephen Plaice and Paul Knight (Cambridge, MA: MIT Press, [1959] 1986), 86–87.

25. Gaston Bachelard, *The Poetics of Reverie*, trans. Daniel Russell (Boston: Beacon Press, [1960] 1971), 102.

CHAPTER 3

1. William James, *Principles of Psychology*, 1 (New York: Henry Holt, [1890] 1918), 601.

2. Ivan P. Pavlov, *Conditioned Reflexes: An Investigation of the Physiological Activity of the Cerebral Cortex*, trans. G. V. Anrep (London: Oxford University Press, [1927] 1946), 266.

3. Ernest Hilgard, *Divided Consciousness: Multiple Controls in Human Thought and Action* (New York: John Wiley & Sons, 1977), 159.

4. Robert Darnton, *Mesmerism and the end of the Enlightenment* (Cambridge, MA: Harvard University Press, 1968), 6.

5. George Makari, *Soul Machine: The Invention of the Modern Mind* (New York: W. W. Norton and Co., 2015), 322–323.

6. Francois Regourd, "Mesmerism in Saint-Domingue," in *Science and Empire in the Atlantic World*, eds. Nicholas Dew and James Delbuorgo (London: Routledge, 2008), 311–332.

7. Johann Heinrich Jung-Stilling, *Theory of Pneumatology* (New York: J. S. Redfield, [1808] 1854), 33–34.

8. Abbé Faria, "On the Cause of Lucid Sleep," trans. S. R. Luis [1819], https://www.abbefaria.com/Sommeil-Luis.htm.

9. Louise Alphonse Cahagnet, *The Celestial Telegraph: Or, Secrets of the Life to Come Revealed Through Magnetism* (New York: J. S. Redfield, 1851), 41.

10. John Elliotson, "Prospectus for the Zoist," in *John Elliotson on Mesmerism*, ed. Fred Kaplan (New York: De Capo Press, [1843] 1983), 286–287.

11. Harriet Martineau, *Letters on Mesmerism* (New York: Harper and Brothers, 1845), 6.

12. Charles Lafontaine, *Memoirs of a Magnetiser*, quoted in *Braid on Hypnotism*, ed. A. W. Waites (New York: Julian Press [1866] 1960), 6–7.

13. James Braid, "Letter to the Lancet," *The Lancet* 45, no. 1135 (May 1845): 627–628.

14. James Esdaile, *Mesmerism in India* (New Delhi: Asian Educational Services, [1846] 1989), 12–14.

15. Axel Munthe, *The Story of San Michele* (London: John Murray, 1929), 302–303.

16. Hippolyte Bernheim, *Suggestive Therapeutics: A Treatise on the Nature and Uses of Hypnotism*, trans. Christian A. Herter (New York: G.P. Putnam's Sons, 1880), VII-IX.

17. Sigmund Freud, "Letter to Wilhelm Fliess, Dec 6, 1896," in *The Complete Letters of Sigmund Freud to Wilhelm Fliess 1887–1904*, trans. and ed. Jeffrey Moussaieff Masson (Cambridge, MA: Harvard University Press, 1985), 207–208.

18. Sigmund Freud, "Group Psychology and the Analysis of Ego," in *The Complete Psychological Works of Sigmund Freud. Volume XVIII*, ed. James Strachey (London: Hogarth Press and Institute of Psychoanalysis, [1921] 1981), 114.

19. Albert Moll, *Hypnotism* (London: W. Scott, 1890), 337–338.

20. James Coates, *How to Mesmerise* (London: W. Foulsham & Co, 1893), 70.

21. F. W. H. Myers, *Human Personality and Its Survival of Bodily Death* (London: Longmans, Green, and Co., 1909), 127–129.

22. Emile Coué, *Self-Mastery Through Conscious Autosuggestion* (Whitefish, Montana: Kessinger Publishing, [1922] 1996), 19–20.

23. George Hoben Estabrooks, *Hypnotism* (New York: E.P. Dutton & Co., [1943] 1946), 185–186.

24. Aldous Huxley, "Letter to George Orwell," (1949), https://archive.org/details/huxley-aldous-letter-to-george-orwell-page-1.

25. Theodore R. Sarbin, "Contributions to Role-Taking Theory: I. Hypnotic Behavior," *Psychological Review* 57, no. 5 (1950): 260.

26. Edward Hunter, *Brain-Washing in Red China: The Calculated Destruction of Men's Minds* (New York: Vanguard Press, 1951), 4.

27. "Subliminal Projection," *Advertising Age* (September 16, 1957), 127.

28. Richard Condon, *The Manchurian Candidate* (New York: New American Library [1959] 1962), 206–207.

29. Martin Gardner, "Bridey Murphy and Other Matters," in *Fads and Fallacies in the Name of Science* (New York: Dover Publications, 1957), 315–316.

30. William S. Kroger and Sidney A. Schneider, "An Electronic Aid for Hypnotic Induction: A Preliminary Report," *International Journal of Clinical and Experimental Hypnosis* 7 no. 2 (1959): 93–98.

31. "Hypnotism and the CIA Operative," *KUBARK Counterintelligence Interrogation* (1963), https://nsarchive2.gwu.edu/NSAEBB/NSAEBB27/docs/doc01.pdf, 96–97.

32. Lynn Schroeder and Sheila Ostrander, *Psychic Discoveries Behind the Iron Curtain* (New York: Bantam Books, 1970), 299–300.

33. Ernest Hilgard, *Divided Consciousness: Multiple Controls in Human Thought and Action* (New York: John Wiley & Sons, 1977), 185–210.

34. John E. Mack, *Abduction: Human Encounters with Aliens* (New York: Ballantine Books, 1995), 53–54.

35. John F. Kihlstrom, "Neuro-Hypnotism: Prospects for Hypnosis and Neuroscience," *Cortex* 49, no. 2 (2013): 365–374.

CHAPTER 4

1. T. H. Huxley, "On the Hypothesis that Animals are Automata," *Collected Essays* (New York: The Macmillan Company, [1874] 1904), 240.

2. Alfred Lord Whitehead, *An Introduction to Mathematics* (New York: Henry Holt, 1911), 60.

3. Eric R. Dodds, *The Greeks and the Irrational* (Berkeley: University of California Press, ([1951] 1973), 57–58.

4. René Descartes, *Meditations on First Philosophy*, trans. Michael Moriarty (Oxford: Oxford University Press, [1641] 2008).

5. Gottfried Leibniz, *New Essays on Human Understanding*, trans. and ed. Peter Remnant and Jonathan Bennett (Cambridge: Cambridge University Press, [1765] 1996), 53–54.

6. Immanuel Kant, *Anthropology from a Pragmatic Point of View*, ed. Manfred Kuehn, trans. Robert B. Louden (Cambridge: Cambridge University Press, ([1798] 2006), 23.

7. Franz Gall, *On the Function of the Brain and Each of its Parts*, trans. Winslow Lewis (Boston: Marsh, Capen & Lyon, [1825] 1835), 274–275.

8. Charles Darwin, "M Notebooks," (Sept 4, 1838), entry 128, http://darwin-online .org.uk.

9. Arthur Schopenhauer, "Some Thoughts Concerning the Intellect," in *Parerga and Paralipomena Vol 2*, ed. and trans. Adrian Del Caro (Cambridge: Cambridge University Press, [1851] 2015), https://doi.org/10.1017/CBO9781139016889.

10. Eneas Sweetland Dallas, *The Gay Science, Vol. 1* (London: Chapman and Hall, 1866), 193–200.

11. Eduard Von Hartmann, *Philosophy of the Unconscious, Vol. 2*, trans. William Chatterton Coupland (London: Kegan Paul, Trench, Trubner & Co., 1884), 39–40.

12. Charles Darwin, *The Expression of Emotions in Man and Animals* (Chicago: University of Chicago, ([1872] 1965), 357–358.

13. William Benjamin Carpenter, *Principles of Mental Physiology* (Cambridge: Cambridge University Press, [1874] 2009), 515.

14. Francis Galton, *Inquiries into Human Faculty and its Development* (London: Macmillan & Co, 1883), 203–204.

15. William James, "What is an Instinct?," *Scriber's Magazine* 1, no.3 (March 1887), 355–366.

16. Joseph Jastrow, *The Subconscious* (Boston: Houghton, Mifflin and Company, 1906), 99–102.

17. Henri Bergson, *Creative Evolution*, trans. Arthur Mitchell (New York: Random House, ([1907] 1911), 182–183.

18. Wolfgang Köhler, *The Mentality of Apes*, trans. Ella Winter (New York: Harcourt, Brace & Company, [1917] 1925), 132–133.

19. Graham Wallas, *The Art of Thought* (London: Jonathan Cape, 1926), 81–82.

20. P. G. Wodehouse, *Right Ho, Jeeves* (Harmondsworth: Penguin, ([1934] 1978), 115.

21. William McDougall, *Introduction to Social Psychology* (London: Methuen & Co., [1936] 1950), 459.

22. Albert Einstein, "Letter to Jacques Hadamard," in Jacques Hadamard, *The Psychology of Invention in the Mathematical Field* (New York: Dover Publications, [1945] 1954), 142–143.

23. Konrad Lorenz, "The Role of Gestalt Perception in Animal and Human Behaviour," in *Aspects of Form*, ed. Lancelot Law Whyte (London: Lund Humphries, 1951), 157–178.

24. Arthur Koestler, "Creativity and the Unconscious," in *Bricks to Babel: Selected Writings* (London: Picador, 1982), 363–364.

25. Noam Chomsky, *Reflections on Language* (New York: Pantheon, 1975), 9–11.

26. David Premack and Guy Woodruff, "Does The Chimpanzee Have a Theory of Mind?" *Behavioral And Brain Sciences* 1, no. 49 (1978): 515–526.

27. Nicholas Humphrey, *The Inner Eye* (London: Faber, 1986), 59–60.

28. Benjamin Libet, *Mind Time: The Temporal Factor in Consciousness* (Cambridge, MA: Harvard University Press, 2004), 129–139.

29. Nancy C. Andreasen, *The Creating Brain: The Neuroscience of Genius* (New York: Dana Press, 2005), 77–78.

CHAPTER 5

1. Saint Augustine, "Psalm 15," in *The Essential Augustine*, ed. Vernon J. Bourke (Indianapolis: Hackett Publishing, [419 A.D] 1974), 40.

2. Socrates, quoted in Plato, *Phaedrus*, in *Complete Works*, ed. J. M. Cooper (Indianapolis: Hackett, [c.399–347 BC] 1997), 551–552.

3. Sigmund Freud, *Civilization and Its Discontents*, trans. Joan Riviere (London: Hogarth Press and The Institute of Psycho-Analysis, 1930), 19.

4. Plato, *Theaetetus*, trans. John McDowell (Oxford: Oxford University Press, [369 BCE] 1973), 79.

5. Aristotle, "On Memory," in *The Complete Works of Aristotle*, ed. Jonathan Barnes (Princeton: Princeton University Press, [350 BCE] 1995), 718.

6. Cicero, *The De Oratore of Cicero*, trans. F. B. Calvert (Edinburgh: Edmonston and Douglas, [55BC] (1870), 141–144.

7. John Locke, *An Essay Concerning Human Understanding*, ed. Roger Woolhouse (Harmondsworth: Penguin, ([1690] 1997), 149.

8. Samuel Johnson, "The Regulation of Memory," in *The Idler* (London: Jones & Company, [1759] 1826), 78.

9. Arthur Schopenhauer, *The World as Will and Representation, Vol. 1* (New York: Dover Publications, [1818] 2000), 400–401.

10. Thomas Chalmers, *The Adaptation of External Nature to the Moral and Intellectual Constitution of Man* (Cambridge: Cambridge University Press ([1835] 2009), 163–164.

11. Thomas De Quincey, *Suspiria De Profundis* (Edinburgh: Adam and Charles Black, ([1845] 1871), 20–21.

12. Théodule-Armand Ribot, *Diseases of Memory* (London: Kegan Paul, Trench & Co., 1882), 127–130.

13. Sigmund Freud, "Letter to Wilhelm Fliess" (Dec 6, 1896), in *The Complete Letters of Sigmund Freud to Wilhelm Fliess 1887–1904*, trans. and ed. Jeffrey Moussaieff Masson (Cambridge, MA: Harvard University Press, 1985), 207–208.

14. Henri Bergson, *Matter and Memory*, trans. Margaret Nancy Paul and W. Scott Palmer (London: George Allen & Unwin Ltd, ([1896] 1911), 90.

15. Theodor Flournoy, *From India to the planet Mars: A Psychic study of a Case of Somnambulism*, trans. D. B. Vermilye (New York: Harper & Brothers, 1900), 275–276.

16. Édouard Claparède, "Recognition And 'Me-Ness,'" in *Organization and Pathology of Thought*, ed. David Rapaport (New York: Columbia University Press, [1911] 1951), 68–70.

17. Frederic C. Bartlett, *Remembering: A Study in Experimental and Social Psychology* (Cambridge: Cambridge University Press, [1932] 1995), 213–214.

18. Carl Gustav Jung, *The Collected Works of C. G. Jung, Vol. 9*, ed. Gerhard Adler, trans. R. F. C. Hull (Princeton: Princeton University Press [1936] 1981), 42–43.

19. Sigmund Freud, *Moses and Monotheism*, trans. Katherine Jones (London: Hogarth Press and the Institute of Psychoanalysis, [1939] 1940), 159–160.

20. L. Ron Hubbard, *The Modern Science of Mental Health* (Commerce, CA: Bridge Publications, [1950] 2007).

21. Alexander Luria, *The Mind of a Mnemonist: A Little Book about a Vast Memory*, trans. Lynn Solotaroff (New York: Basic Books, 1968), 34–35.

22. Wilder Penfield, *Mystery of the Mind* (Princeton: Princeton University Press, 1975), 21–22.

23. Howard Gardner, *The Shattered Mind* (London: Routledge & Kegan Paul, 1977), 183–186.

24. Marvin Minsky, *Society of Mind* (New York: Simon and Schuster, 1986), 154.

25. Endel Tulving and Daniel L. Schacter, "Priming and Human Memory Systems," *Science* 247, no. 4940 (February 1990): 301–306.

26. Elizabeth F. Loftus, "Creating False Memories," *Scientific American* 277, no. 3 (September 1997), 70–75.

CHAPTER 6

1. Thomas Reid, *Essays on the Active Powers of Man* (Edinburgh: John Bell, 1788), 119.

2. William James, *Principles of Psychology*, 1 (New York: Henry Holt, [1890] 1918), 112.

3. John B. Watson, *Behaviorism* (New York: W. W. Norton, [1924] 1970), 104.

4. Grace Adams, "The Rise and Fall of Behaviourism," *The Atlantic Monthly* (Jan 1, 1934).

5. John Locke, *Of the Conduct of the Understanding* (Oxford: Clarendon Press, [1706] 1881), 13–14.

6. David Hume, *An Enquiry Concerning Human Understanding*, ed. Peter Millican (Oxford: Oxford University Press, [1748] 2007), 31–33.

7. Adam Smith, *An Inquiry into the Nature and Causes of the Wealth of Nations* (1776), Part III, https://www.gutenberg.org/files/3300/3300-h/3300-h.htm.

8. Thomas Reid, "Of Power," reprinted in *Philosophical Quarterly* 51, no. 202 ([1792] January 2001): 3.

9. Immanuel Kant, *Anthropology from a Pragmatic Point of View*, ed. Manfred Kuehn, trans. Robert B. Louden (Cambridge: Cambridge University Press, ([1798] 2006), 34–35.

10. Maria Edgeworth and Richard Lovell Edgeworth, *Practical Education* (London: J. Johnson, 1798), 250–251.

11. Jean-Baptiste Lamarck, *Zoological Philosophy: An Exposition with Regard the Natural History of Animals*, trans. Hugh Elliot (Chicago: University of Chicago Press, [1809] 1989), 113–115.

12. Felix Ravaisson, *Of Habit*, trans. Clare Carlisle and Mark Sinclair (London: Continuum [1838] 2008), 57–59.

13. Herbert Spencer, *Principles of Psychology* (London: Longman, Brown, Green, and Longman, 1855), 526–527.

14. Samuel Butler, *Life and Habit* (London: Trubner & Co., 1878), 294–295.

15. Friedrich Nietzsche, *The Gay Science*, trans. Josefine Nauckhoff (Cambridge: Cambridge University Press ([1882] 2007), 167–168.

16. William James, *Psychology: Briefer Course* (New York: Henry Holt and Company, 1892), 144–145.

17. Edward L. Thorndike, *The Principles of Teaching: Based on Psychology* (New York: Mason Press, 1906), 110–111.

18. Ivan P. Pavlov, *Conditioned Reflexes: An Investigation of the Physiological Activity of the Cerebral Cortex*, trans. G. V. Anrep (London: Oxford University Press, [1927] 1946), 24–25.

19. John B. Watson, "Psychology as the Behaviorist Views It," *Psychological Review*, 20 (1913), 158–177.

20. Emil Kraepelin, *Dementia Praecox and Paraphrenia*, trans. Mary R. Barclay (Chicago: Chicago Medical Book Company, [1909] 1919), 44–45.

21. John Dewey, *Human Nature and Conduct* (New York: Henry Holt and Company, 1922), 75–76.

22. Italo Svevo, *Zeno's Conscience*, trans. William Weaver (London: Penguin, [1923] 2002), 12.

23. F. M. Alexander, *Constructive Conscious Control of the Individual* (Bexley, Kent: Integral Press, [1923] 1955), 222–238.

24. Jean Piaget, *The Origins of Intelligence in Children* (New York: International Universities Press, [1936] 1956), 122–137.

25. Maurice Merleau-Ponty, *The Phenomenology of Perception*, trans. D. A. Landes (London: Routledge, [1945] 2012), 144–146.

26. Gilbert Ryle, *The Concept of Mind* (London: Routledge, [1949] 2009), 85.

27. Raymond Ruyer, *Neofinalism*, trans. Alyosha Edlebi (Minneapolis: University of Minnesota Press [1952] 2016), 214–215.

28. Claude Lévi-Strauss, *Structural Anthropology*, trans. Claire Jacobson (New York: Basic Books [1958] 1963), 18.

29. Ellen Langer, Arthur Blank and Benzion Chanowitz, "The Mindlessness of Ostensibly Thoughtful Action," *Journal of Personality and Social Psychology* 36, no. 6 (1978): 635–641.

30. John Bargh, *Before You Know It: The Unconscious Reasons We Do What We Do* (London: William Heinemann, 2017), 178–181.

31. Wendy Wood, *Good Habits, Bad Habits* (London: Macmillan, 2019), 161–162.

CHAPTER 7

1. James Hogg, *The Private Memoirs and Confessions of a Justified Sinner* (London: Longman, 1824), 303.

2. William McDougall, quoted in Michael C. Corballis and Paul M. Corballis, "Can the Mind Be Split? A Historical Introduction," *Neuropsychologia* 163 (2021): 108041, https://doi.org/10.1016/j.neuropsychologia.2021.108041.

3. Roger W. Sperry, "Hemisphere Deconnection and Unity in Conscious Awareness," *American Psychologist* 23, no. 7 (1968): 723–33.

4. Henry Dewar, "Report on a Communication from Dr Dyce of Aberdeen, to the Royal Society of Edinburgh, 'On Uterine Irritation, And its Effects on the Female Constitution'," *Transactions of the Royal Society of Edinburgh* 9, no. 2, (1823): 365–379.

5. Silas Weir Mitchell, *Mary Reynolds: A Case of Double Consciousness* (Philadelphia: William J. Dornan, 1889), 12–13.

6. Arthur Ladbroke Wigan, *The Duality of the Mind: A New View of Insanity* (London: Longman, Brown, Green and Longmans, 1844), 25–27.

7. Francis Wayland, *Elements of Intellectual Philosophy* (New York: Sheldon and Company, [1854] 1869), 115–116.

8. Sigmund Freud and Josef Breuer, *Studies in Hysteria*, trans. James and Alix Strachey (Harmondsworth: Penguin, [1893] 2004), 48–50.

9. Robert Louis Stevenson, *The Strange Case of Dr Jekyll and Mr Hyde* (Scotts Valley, CA: CreateSpace Independent Publishing Platform, [1896] 2016), 84–85.

10. W. E. B. Du Bois, *The Souls of Black Folk* (Oxford: Oxford University Press, [1903] 2007), 8–9.

11. Eugene Bleuler, *Dementia Praecox or the Group of Schizophrenias* (New York: International Universities Press, ([1911] 1950), 130–159.

12. Karl Jaspers, *General Psychopathology* (Baltimore: Johns Hopkins University Press, [1913] 1997), 381–382.

13. Charles E. Cory, "Patience Worth," *Psychological Review* 26, no. 5 (1919): 397–406.

14. Sigmund Freud, "The Ego and the Id," in *The Complete Psychological Works of Sigmund Freud, Volume XI*, trans. Joan Riviere (London: Hogarth Press and Institute of Psychoanalysis (1923–1925), 24–25.

15. Anna Freud, *The Ego and the Mechanisms of Defence*, trans. Cecil Baines (New York: International Universities Press, [1936] 1966), 7.

16. Melanie Klein, "Notes on Some Schizoid Mechanisms," *International Journal of Psycho-Analysis* 27 (1946): 99–110.

17. George Orwell, *Nineteen Eighty-Four* (London: HarperCollins, [1948] 2021), 37.

18. Louis N. Gould, "Verbal Hallucinations as Automatic Speech," *The American Journal of Psychiatry* 107, no. 2 (1950): 110–119.

19. R. D. Laing, *The Divided Self* (Harmondsworth: Penguin [1960] 1984), 152–153.

20. Henri Michaux, *The Major Ordeals of the Mind*, trans. Richard Howard (New York: Harcourt Brace Jovanovich, [1966] 1974), 77.

21. Donald W. Winnicott, *Playing and Reality* (London: Routledge, ([1971] 1991), 29–30.

22. Gilles Deleuze and Felix Guattari, *Anti-Oedipus: Capitalism and Schizophrenia* (Minneapolis: University of Minnesota ([1972] 2000), 245.

23. Julian Jaynes, *The Origin of Consciousness in the Breakdown of The Bicameral Mind* (Harmondsworth: Penguin Books, ([1977] 1993), 72–74.

24. Joseph LeDoux, Donald H. Wilson, and Michael S. Gazzaniga, "A Divided Mind: Observations on the Conscious Properties of the Separated Hemispheres," *Annals of Neurology* 2, no. 5 (1977): 417–421.

25. Philip K. Dick, *A Scanner Darkly* (London: Millenium, [1977] 1999), 167–168.

26. *Diagnostic and Statistical Manual of Mental Disorders*, 3rd edition (Washington: American Psychiatric Association, 1980), 257.

27. Corbett H. Thigpen and Hervey M. Cleckley, "On the Incidence of Multiple Personality Disorder: A Brief Communication," *International Journal of Clinical and Experimental Hypnosis* 32, no. 2 (1984): 63–66.

CHAPTER 8

1. Max Weber, *Max Weber's Science as a Vocation*, ed. Peter Lasman and Peter Velody (London: Unwin Hyman [1918] 1989), 30.

2. Ralph Waldo Emerson, "The Over-Soul," in *The Complete Essays and Other Writings of Ralph Waldo Emerson*, ed. Brooks Atkinson (New York: Random House, [1841] 1950).

3. Robert C. Fuller, *Americans and the Unconscious* (New York: Oxford University Press, 1995).

4. Helena Blavatsky, *The Secret Doctrine* (London: The Theosophical Publishing Company, 1888).

5. Edward Carpenter, "A Consciousness Without Thought," in *From Adam's Peak to Elephanta: Sketches in Ceylon and India* (London: Swan Sonnenschein & Co., 1892).

6. F. W. H. Myers, "The Subliminal Consciousness," in *Proceedings of the Society for Psychical Research* 7 (1982).

7. Swami Vivekananda, *Raja Yoga* (Leeds: Celephaïs Press, [1896] 2003).

8. Annie Besant, *Study in Consciousness: A Contribution to the Science* (London: Theosophical Publishing Society, [1947] 1915).

9. D. H. Lawrence, *Psychoanalysis and the Unconscious and Fantasia of the Unconscious*, ed. Bruce Steele (Cambridge: Cambridge University Press, [1921] 2004).

10. Mable Dodge Luhan, *Intimate Memories: The Autobiography of Mabel Dodge Luhan*, ed. Lois Palken Rudnick, Volume Four (Albuquerque: University of New Mexico Press, [1937] 1999).

11. D. T. Suzuki, *An Introduction to Zen Buddhism* (New York: Philosophical Library, [1934] 1949).

12. Carl Gustav Jung, *Aion: Researches into the Phenomenology of the Self* (New York: Pantheon Books, [1951] 1959.)

13. Joseph Banks Rhine, *New World of The Mind* (New York: William Sloane, 1953).

14. P. D. Ouspensky, *The Fourth Way* (London: Routledge Kegan Paul, [1957] 1967).

15. Aldous Huxley, *The Doors of Perception and Heaven and Hell* (London: Panther Books, [1954] 1985).

16. Teilhard de Chardin, *The Phenomenon of Man* (New York: Harper Perennial, [1955] 2008).

17. Mircea Eliade, *Yoga: Immortality and Freedom* (New York: Pantheon Books, 1958).

18. Alan Watts, *The Book: On the Taboo Against Knowing Who You Are* (London: Abacus, [1966] 1973).

19. Fritz Perls, *The Gestalt Approach and Eyewitness to Therapy* (New York: Bantam Books, [1973] 1981).

20. Theodore Roszak, *The Making of a Counter Culture: Reflections on the Technocratic Society and Its Youthful Opposition* (London: Faber, [1969] 1970).

21. Gay Luce and Erik Peper, "Mind Over Body, Mind Over Mind," *New York Times Magazine*, Sept 12, 1971.

22. Michael Grosso, "The Myth of the Near-Death Journey," *Journal of Near-Death Studies* 10, no.1 (1991): 49–60.

23. Ervin Laszlo, *Science and the Akashic Field: An Integral Theory of Everything* (Rochester, VT: Inner Traditions, 2004).

EPILOGUE

1. Anne Harrington, quoted in James Davies, *Cracked: Why Psychiatry is Doing More Harm Than Good* (London: Icon Books, 2013), 238.

BIBLIOGRAPHY

Alexander, F. M. *Constructive Conscious Control of the Individual*. Bexley, Kent: Integral Press, [1923] 1955.

Andreasen, Nancy C. *The Creating Brain: The Neuroscience of Genius*. New York: Dana Press, 2005.

Aristotle. "On Memory." In *The Complete Works of Aristotle*, ed. Jonathan Barnes. Princeton: Princeton University Press, [350 BCE] 1995, 718.

Arnold-Forster, Mary. *Studies in Dreams*. London: George Allen and Unwin, 1921.

Artemidorus. *Artemidorus' Oneirocritica: Text, Translation, and Commentary*. Translated by Daniel Harris-McCoy. Oxford: Oxford University Press, [2nd Century AD] 2012.

Bachelard, Gaston. *The Poetics of Reverie*. Translated by Daniel Russell. Boston: Beacon Press, [1960] 1971.

Bargh, John. *Before You Know It: The Unconscious Reasons We Do What We Do*. London: William Heinemann, 2017.

Bartlett, Frederic C. *Remembering: A Study in Experimental and Social Psychology*. Cambridge: Cambridge University Press, [1932] 1995.

Bergson, Henri. *Creative Evolution*. Translated by Arthur Mitchell. New York: Random House, [1907] 1911.

Bergson, Henri. *Dreams*. Translated by Edwin E Slosson. New York: B. W. Huebsch, 1914.

Bergson, Henri. *Matter and Memory*. Translated by Margaret Nancy Paul and W. Scott Palmer. London: George Allen & Unwin Ltd, [1896] 1911.

Bernheim, Hippolyte. *Suggestive Therapeutics: A Treatise on the Nature and Uses of Hypnotism*. Translated by Christian A. Herter. New York: G.P. Putnam's Sons, 1880.

Besant, Annie. *Study in Consciousness: A Contribution to the Science*. London: Theosophical Publishing Society, [1947] 1915.

Blavatsky, Helena. *The Secret Doctrine*. London: The Theosophical Publishing Company, 1888.

Bleuler, Eugene. *Dementia Praecox or the Group of Schizophrenias*. New York: International Universities Press, [1911] 1950.

Bloch, Ernst. *The Principle of Hope, Vol. 1*. Translated by Neville Plaice, Stephen Plaice and Paul Knight. Cambridge, MA: MIT Press, [1959] 1986.

Braid, James. "Letter to the Lancet." *The Lancet* 45, no. 1135 (May 1845): 627–628.

Breton, André. *What Is Surrealism?* Edited by Franklin Rosemont. Translated by David Gascoyne. New York: Monad Press, [1934] 1978.

Butler, Samuel. *Life and Habit*. London: Trubner & Co., 1878.

Cahagnet, Louise Alphonse. *The Celestial Telegraph: Or, Secrets of the Life to Come Revealed Through Magnetism*. New York: J. S. Redfield, 1851.

Carpenter, Edward. "A Consciousness Without Thought." In *From Adam's Peak to Elephanta: Sketches in Ceylon and India*. London: Swan Sonnenschein & Co., 1892.

Carpenter, William Benjamin. *Principles of Mental Physiology*. Cambridge: Cambridge University Press, [1874] 2009.

Chalmers, Thomas. *The Adaptation of External Nature to the Moral and Intellectual Constitution of Man*. Cambridge: Cambridge University Press [1835] 2009.

Chomsky, Noam. *Reflections on Language*. New York: Pantheon, 1975.

Cicero. *The De Oratore of Cicero*. Translated by F. B. Calvert. Edinburgh: Edmonston and Douglas, [55BC] 1870, 141–144.

Cioran, E. M. *On the Heights of Despair*. Translated by Ilinca Zarifopol-Johnston. Chicago: University of Chicago Press, [1934] 1992.

Claparède, Édouard. "Recognition And 'Me-Ness." *Organization and Pathology of Thought*, ed. David Rapaport, 68–70. New York: Columbia University Press, [1911] 1951.

Coates, James. *How to Mesmerise*. London: W. Foulsham & Co, 1893.

Cobbe, Frances Power. "Dreams as Illustration of Involuntary Cerebration." In *Darwinism in Morals and Other Essays*. London: William and Norgate, 1872.

Condon, Richard. *The Manchurian Candidate*. New York: New American Library [1959] 1962.

Cory, Charles E. "Patience Worth." *Psychological Review* 26, no. 5 (1919): 397–406.

Coué, Emile. *Self-Mastery Through Conscious Autosuggestion*. Whitefish, Montana: Kessinger Publishing, [1922] 1996.

Crick, Francis, and Graeme Mitchison. "The Function of Dream Sleep." *Nature* 304, no. 5922, (July 1983): 111–114.

Cudworth, Ralph. *A Treatise Concerning Eternal and Immutable Morality*. Edited by Sarah Hutton. Cambridge: Cambridge University Press, [1731] 1996.

Dallas, Eneas Sweetland. *The Gay Science, Vol. 1*. London: Chapman and Hall, 1866.

Darnton, Robert. *Mesmerism and the end of the Enlightenment*. Cambridge, MA: Harvard University Press, 1968.

Darwin, Charles. "M Notebooks." Sept 4, 1838, entry 128.

Darwin, Charles. *The Expression of Emotions in Man and Animals*. Chicago: University of Chicago, [1872] 1965.

Darwin, Erasmus. *Zoonomia, Vol. 1*. Boston: Thomas and Andrews, 1803.

de Cervantes, Miguel. *Don Quixote De La Mancha*. Translated by Charles Jarvis. Oxford: Oxford University Press, [1605] 2008.

de Chardin, Teilhard. *The Phenomenon of Man*. New York: Harper Perennial, [1955] 2008.

de Manaciene, Marie. *Sleep: Its Physiology, Pathology, Hygiene, and Psychology*. London: Walter Scott Ltd, 1897.

Deleuze, Gilles, and Felix Guattari. *Anti-Oedipus: Capitalism and Schizophrenia*. Minneapolis: University of Minnesota [1972] 2000.

Dement, William C. *Some Must Watch While Some Must Sleep*. New York: Norton, 1974.

Descartes, René. *Meditations on First Philosophy*. Translated by Michael Moriarty. Oxford: Oxford University Press, [1641] 2008.

Dewar, Henry. "Report on a Communication from Dr Dyce of Aberdeen, to the Royal Society of Edinburgh, 'On Uterine Irritation, And its Effects on the Female Constitution'." *Transactions of the Royal Society of Edinburgh* 9, no.2, (1823).

Dewey, John. *Human Nature and Conduct*. New York: Henry Holt and Company, 1922.

Diagnostic and Statistical Manual of Mental Disorders. 3rd edition. Washington: American Psychiatric Association, 1980.

Dick, Philip K. *A Scanner Darkly*. London: Millenium, [1977] 1999.

Dodds, Eric R. *The Greeks and the Irrational*. Berkeley: University of California Press, [1951] 1973.

De Quincey, Thomas. *Suspiria De Profundis*. Edinburgh: Adam and Charles Black, [1845] 1871.

Du Bois, W. E. B. *The Souls of Black Folk*. Oxford: Oxford University Press, [1903] 2007.

Edgeworth, Maria, and Richard Lovell Edgeworth. *Practical Education*. London: J. Johnson, 1798.

Einstein, Albert. "Letter to Jacques Hadamard." Jacques Hadamard, *The Psychology of Invention in the Mathematical Field*. New York: Dover Publications, [1945] 1954.

Ekirch, Roger A. *At Day's Close: A History of Nighttime*. New York: Norton, 2005.

Eliade, Mircea. *Yoga: Immortality and Freedom*. New York: Pantheon Books, 1958.

Ellis, Havelock. *The World of Dreams*. Boston: Houghton Mifflin Company [1911] 1922.

Elliotson, John. "Prospectus for the Zoist." *John Elliotson on Mesmerism*. Edited Fred Kaplan. New York: De Capo Press, [1843] 1983.

Emerson, Ralph Waldo. "The Over-Soul." In *The Complete Essays and Other Writings of Ralph Waldo Emerson*. Edited by Brooks Atkinson. New York: Random House [1841] 1950.

Estabrooks, George Hoben. *Hypnotism*. New York: E.P. Dutton & Co., [1943] 1946.

Esdaile, James. *Mesmerism in India*. New Delhi: Asian Educational Services, [1846] 1989.

Faria, Abbé. "On the Cause of Lucid Sleep." Translated by Luis S. R. Vas, [1819] 2025, https://www.abbefaria.com/Sommeil-Luis.htm.

Flournoy, Theodore. *From India to the planet Mars: A Psychic study of a Case of Somnambulism*. Translated by D. B. Vermilye. New York: Harper & Brothers, 1900.

Foulkes, David. *Children's Dreaming and The Development of Consciousness*. Cambridge, MA: Harvard University Press, 1999.

Freud, Sigmund. *Moses and Monotheism*. Translated by Katherine Jones. London: Hogarth Press and the Institute of Psychoanalysis, [1939] 1940.

Freud, Anna. *The Ego and the Mechanisms of Defence*. Translated by Cecil Baines. New York: International Universities Press, [1936] 1966.

Freud, Sigmund. "Group Psychology and the Analysis of Ego." In *The Complete Psychological Works of Sigmund Freud. Volume XVIII*, edited by James Strachey. London: Hogarth Press and Institute of Psychoanalysis, [1921] 1981.

Freud, Sigmund. Letter to Wilhelm Fliess (Dec 6, 1896). In *The Complete Letters of Sigmund Freud to Wilhelm Fliess 1887–1904*, translated and edited by Jeffrey Moussaieff Masson. Cambridge, MA: Harvard University Press, 1985.

Freud, Sigmund. "The Ego and the Id." In *The Complete Psychological Works of Sigmund Freud, Volume XI*, translated by Joan Riviere. London: Hogarth Press and Institute of Psychoanalysis, 1923–1925.

Freud, Sigmund. *The Interpretation of Dreams: The Complete and Definitive Text*. Edited and translated by James Strachey. Basic Books: New York [1900] 2010.

Freud, Sigmund, and Josef Breuer. *Studies in Hysteria*. Translated by James and Alix Strachey. Harmondsworth: Penguin, [1893] 2004.

Fuller, Robert C. *Americans and the Unconscious*. New York: Oxford University Press, 1995.

Gall, Franz. *On the Function of the Brain and Each of its Parts*. Translated by Winslow Lewis Boston: Marsh, Capen & Lyon, [1825] 1835.

Galton, Francis. *Inquiries into Human Faculty and its Development*. London: Macmillan & Co, 1883.

Gardner, Howard. *The Shattered Mind*. London: Routledge & Kegan Paul, 1977.

Gardner, Martin. "Bridey Murphy and Other Matters." *Fads and Fallacies in the Name of Science*. New York: Dover Publications (1957), 315–316.

Godwin, William. *Thoughts on Man*. London: Effingham Wilson, 1831.

Grosso, Michael. "The Myth of the Near-Death Journey." *Journal of Near-Death Studies* 10, no.1 (1991): 49–60.

Gould, Louis N. "Verbal Hallucinations as Automatic Speech." *The American Journal of Psychiatry* 107, no. 2 (1950): 110–119.

Hall, Calvin S. "What People Dream About." *Scientific American* 184, no. 5 (May 1951).

Hartmann, Eduard Von. *Philosophy of the Unconscious, Vol. 2*. Translated by William Chatterton Coupland. London: Kegan Paul, Trench, Trubner & Co., 1884.

Hazlitt, William. "On Dreams." In *The Plain Speaker, Volume 1*. London: Henry Colburn, 1826.

Hervey de Saint-Denys, Leon. *Dreams and How to Guide Them*. Translated by Peter Connor. Edited by Phil Baker. London: Strange Attractor Press, 2025.

Hilgard, Ernest. *Divided Consciousness: Multiple Controls in Human Thought and Action*. New York: John Wiley & Sons, 1977.

Hippocrates. *Hippocrates, Volume IV*. Translated by W. H. S. Jones. Cambridge, MA: Harvard University Press, [400 BC] 1931.

Hobbes, Thomas. *Leviathan*. Edited by J. C. A. Gaskin. Oxford: Oxford University Press, [1651] 1998.

Hobson, J. Allan. *Dream Life: An Experimental Memoir*. Cambridge, MA: MIT Press, 2011.

Howells, William Dean. "True, I talk of Dreams." In *Oxford Book of American Essays*, edited by by Brander Matthews. New York: Oxford University Press, [1895] 1914.

Hubbard, L. Ron. *The Modern Science of Mental Health*. Commerce, CA: Bridge Publications, [1950] 2007.

Hume, David. *An Enquiry Concerning Human Understanding*. Edited by Peter Millican. Oxford: Oxford University Press, [1748] 2007.

Humphrey, Nicholas. *The Inner Eye*. London: Faber, 1986.

Hunter, Edward. *Brain-Washing in Red China: The Calculated Destruction of Men's Minds*. New York: Vanguard Press, 1951.

Huxley, Aldous. *The Doors of Perception and Heaven and Hell*. London: Panther Books, [1954] 1985.

Huxley, Aldous. "Letter to George Orwell." (1949), https://archive.org/details/huxley -aldous-letter-to-george-orwell-page-1.

"Hypnotism and the CIA Operative." *KUBARK Counterintelligence Interrogation*. (1963), https://nsarchive2.gwu.edu/NSAEBB/NSAEBB27/docs/doc01.pdf.

James, William. *Psychology: Briefer Course*. New York: Henry Holt and Company, 1892.

James, William. "What is an Instinct?" *Scriber's Magazine* 1, no. 3 (March 1887): 355–366.

Janet, Pierre. *The Mental State of Hystericals*. Edited by Caroline Rollin Corson. New York: G. P. Putnam, [1892] 1901.

Jaspers, Karl. *General Psychopathology*. Baltimore: Johns Hopkins University Press, [1913] 1997.

Jastrow, Joseph. *The Subconscious*. Boston: Houghton, Mifflin and Company, 1906.

Jaynes, Julian. *The Origin of Consciousness in the Breakdown of The Bicameral Mind*. Harmondsworth: Penguin Books, [1977] 1993.

Johnson, Samuel. "The Regulation of Memory." *The Idler*. London: Jones & Company, [1759] 1826.

Jouvet, Michel. *The Paradox of Sleep*. Cambridge, MA: MIT Press, 1999.

Jung, Carl Gustav. *Aion: Researches into the Phenomenology of the Self*. New York: Pantheon Books, [1951] 1959.

Jung, Carl Gustav. *The Collected Works of C. G. Jung, Vol. 9*. Edited by Gerhard Adler. Translated by R. F. C. Hull. Princeton: Princeton University Press, [1936] 1981.

Jung-Stilling, Johann Heinrich. *Theory of Pneumatology*. New York: J. S. Redfield, [1808] 1854.

Kant, Immanuel. *Anthropology from a Pragmatic Point of View*. Edited by Manfred Kuehn. Translated by Robert B. Louden. Cambridge: Cambridge University Press, [1798] 2006.

Kihlstrom, John F. "Neuro-Hypnotism: Prospects for Hypnosis and Neuroscience." *Cortex* 49, no. 2 (2013): 365–374.

Klein, Melanie. "Notes on Some Schizoid Mechanisms." *International Journal of Psycho-Analysis* 27 (1946).

Koestler, Arthur. "Creativity and the Unconscious." In *Bricks to Babel: Selected Writings*. London: Picador, 1982.

Köhler, Wolfgang. *The Mentality of Apes*. Translated by Ella Winter. New York: Harcourt, Brace & Company, [1917] 1925.

Kraepelin, Emil. *Dementia Praecox and Paraphrenia*. Translated by Mary R. Barclay. Chicago: Chicago Medical Book Company, [1909] 1919.

Kroger, William S., and Sidney A. Schneider. "An Electronic Aid for Hypnotic Induction: A Preliminary Report." *International Journal of Clinical and Experimental Hypnosis* 7, no. 2 (1959): 93–98.

Lafontaine, Charles. *Memoirs of a Magnetiser*. Quoted in *Braid on Hypnotism*, edited by A. W. Waites. New York: Julian Press, [1866] 1960.

Laing, R. D. *The Divided Self*. Harmondsworth: Penguin, [1960] 1984.

Lamarck, Jean-Baptiste. *Zoological Philosophy: An Exposition with Regard the Natural History of Animals*. Translated by Hugh Elliot. Chicago: University of Chicago Press, [1809] 1989.

Langer, Ellen, Arthur Blank, and Benzion Chanowitz. "The Mindlessness of Ostensibly Thoughtful Action." *Journal of Personality and Social Psychology* 36, no. 6 (1978): 635–641.

Lawrence, D. H. *Psychoanalysis and the Unconscious and Fantasia of the Unconscious*. Edited by Bruce Steele. Cambridge: Cambridge University Press, [1921] 2004.

Laszlo, Ervin. *Science and the Akashic Field: An Integral Theory of Everything*. Rochester: Inner Traditions, 2004.

Leader, Darian. *Why Can't We Sleep*. London: Hamish Hamilton, 2019.

LeDoux, Joseph, Donald H. Wilson, and Michael S. Gazzaniga. "A Divided Mind: Observations on the Conscious Properties of the Separated Hemispheres," *Annals of Neurology* 2, no. 5 1977.

Leibniz, Gottfried. *New Essays on Human Understanding*. Translated and edited by Peter Remnant and Jonathan Bennett. Cambridge: Cambridge University Press, [1765] 1996.

Lévi-Strauss, Claude. *Structural Anthropology*. Translated by Claire Jacobson. New York: Basic Books, [1958] 1963.

Libet, Benjamin. *Mind Time: The Temporal Factor in Consciousness*. Cambridge, MA: Harvard University Press, 2004.

Locke, John. *An Essay Concerning Human Understanding*. Edited by Roger Woolhouse. Harmondsworth: Penguin, [1690] 1997.

Loftus, Elizabeth F. "Creating False Memories." *Scientific American* 277, no. 3 (September 1997): 70–75.

Lorenz, Konrad. "The Role of Gestalt Perception in Animal and Human Behaviour." *Aspects of Form*. Edited by Lancelot Law Whyte. London: Lund Humphries, 1951.

Luce, Gay, and Erik Peper. "Mind Over Body, Mind Over Mind." *New York Times Magazine*, Sept 12, 1971.

Luhan, Mable Dodge. *Intimate Memories: The Autobiography of Mabel Dodge Luhan*. Edited by Lois Palken Rudnick. Volume four. Albuquerque: University of New Mexico Press, [1937] 1999.

Locke, John. *Of the Conduct of the Understanding*. Oxford: Clarendon Press, [1706] 1881.

Luria, Alexander. *The Mind of a Mnemonist: A Little Book about a Vast Memory*. Translated by Lynn Solotaroff. New York: Basic Books, 1968.

Mack, John E. *Abduction: Human Encounters with Aliens*. New York: Ballantine Books, 1995.

Mackay, Charles. *Extraordinary Popular Delusions and The Madness of Crowds*. London: Wordsworth Books, [1841] 1995.

Makari, George. *Soul Machine: The Invention of the Modern Mind*. New York: W. W. Norton and Co., 2015.

Martineau, Harriet. *Letters on Mesmerism*. New York: Harper and Brothers, 1845.

McDougall, William. *Introduction to Social Psychology*. London: Methuen & Co., [1936] 1950.

Merleau-Ponty, Maurice. *The Phenomenology of Perception*. Translated by D. A. Landes. London: Routledge, [1945] 2012.

Michaux, Henri. *The Major Ordeals of the Mind*. Translated by Richard Howard. New York: Harcourt Brace Jovanovich, [1966] 1974.

Minsky, Marvin. *Society of Mind*. New York: Simon and Schuster, 1986.

Mitchell, Silas Weir. *Mary Reynolds: A Case of Double Consciousness*. Philadelphia: William J. Dornan, 1889.

Moll, Albert. *Hypnotism*. London: W. Scott, 1890.

Montaigne, Michel. *The Complete Works of Montaigne*. Translated by D. M. Frame. Stanford: Stanford University Press, [1580 BC] 1957.

Moreau, Jacques-Joseph. *Hashish and Mental Illness*. Translated by Gordon J. Barnett. New York: Raven Press, [1845] 1975.

Munthe, Axel. *The Story of San Michele*. London: John Murray, 1929.

Myers, F.W.H. *Human Personality and Its Survival of Bodily Death*. London: Longmans, Green, and Co., 1909.

Myers, F.W.H. "The Subliminal Consciousness." In *Proceedings of the Society for Psychical Research* 7, 1982.

Nabokov, Vladimir. *Speak, Memory*. New York: Vintage Books, [1947] 1989.

Nietzsche, Friedrich. *Human, All Too Human*. Translated by R. J. Hollingdale. Cambridge: Cambridge University Press, [1878] 1996.

Nietzsche, Friedrich. *The Gay Science*. Translated by Josefine Nauckhoff. Cambridge: Cambridge University Press, [1882] 2007.

Orwell, George. *Nineteen Eighty-Four*. London: HarperCollins, [1948] 2021.

Ouspensky, P.D. *The Fourth Way*. London: Routledge Kegan Paul, [1957] 1967.

Pavlov, Ivan P. *Conditioned Reflexes: An Investigation of the Physiological Activity of the Cerebral Cortex*. Translated by G. V. Anrep. London: Oxford University Press, [1927] 1946.

Penfield, Wilder. *Mystery of the Mind*. Princeton: Princeton University Press, 1975.

Perls, Fritz. *The Gestalt Approach and Eyewitness to Therapy*. New York: Bantam Books, [1973] 1981.

Piaget, Jean. *The Origins of Intelligence in Children*. New York: International Universities Press, [1936] 1956.

Plato. *Theaetetus*. Translated by John McDowell. Oxford: Oxford University Press, [369 BCE] 1973, 79.

Premack, David, and Guy Woodruff. "Does The Chimpanzee Have a Theory of Mind?" *Behavioral And Brain Sciences* 1, no. 49 (1978): 515–526.

Ravaisson, Felix. *Of Habit*. Translated by Clare Carlisle and Mark Sinclair. London: Continuum, [1838] 2008.

Regourd, Francois. "Mesmerism in Saint-Domingue." In *Science and Empire in the Atlantic World*, edited by Nicholas Dew and James Delbuorgo. London: Routledge, 2008.

Reid, Thomas. "Of Power." Reprinted in *Philosophical Quarterly* 51, no. 202 ([1792] January 2001): 3.

Rhine, Joseph Banks. *New World of The Mind*. New York: William Sloane, 1953.

Ribot, Théodule-Armand. *Diseases of Memory*. London: Kegan Paul, Trench & Co., 1882.

Rousseau, Jean-Jacques. *Reveries of the Solitary Walker*. Translated by Russell Gouldbourne. Oxford: Oxford University Press, [1782] 2011.

Roszak, Theodore. *The Making of a Counter Culture: Reflections on the Technocratic Society and Its Youthful Opposition*. London: Faber, [1969] 1970.

Ruyer, Raymond. *Neofinalism*. Translated by Alyosha Edlebi. Minneapolis: University of Minnesota Press, [1952] 2016.

Ryle, Gilbert. *The Concept of Mind*. London: Routledge, [1949] 2009.

Sarbin, Theodore R. "Contributions to Role-Taking Theory: I. Hypnotic Behavior." *Psychological Review* 57, no. 5 (1950): 260.

Schroeder, Lynn, and Sheila Ostrander. *Psychic Discoveries Behind the Iron Curtain*. New York: Bantam Books, 1970.

Schopenhauer, Arthur. "Some Thoughts Concerning the Intellect." *Parerga and Paralipomena Vol 2*. Edited and translated by Adrian Del Caro. Cambridge: Cambridge University Press, [1851] 2015: https://doi.org/10.1017/CBO9781139016889..

Schopenhauer, Arthur. *The World as Will and Representation, Vol. 1*. New York: Dover Publications, [1818] 2000.

Silberer, Herbert. "Report on a Method of Eliciting and Observing Certain Symbolic-Hallucination Phenomena." In *Organization and Pathology of Thought*, edited and translated by David Rapaport. New York: Columbia University Press, [1909] 1951.

Smith, Adam. *An Inquiry into the Nature and Causes of the Wealth of Nations*. Part III, 1776, https://www.gutenberg.org/files/3300/3300-h/3300-h.htm.

Smith, Theodate L. "The Psychology Of Day Dreams." *The American Journal of Psychology* 15, no. 4, October 1904.

Spencer, Herbert. *Principles of Psychology*. London: Longman, Brown, Green, and Longman, 1855.

Stevenson, Robert Louis. *The Strange Case of Dr Jekyll and Mr Hyde*. Scotts Valley, CA: CreateSpace Independent Publishing Platform, [1896] 2016.

Suzuki, D.T. *An Introduction to Zen Buddhism*. New York: Philosophical Library, [1934] 1949.

Svevo, Italo. *Zeno's Conscience*. Translated by William Weaver. London: Penguin, [1923] 2002.

Synesius of Cyrene. *The Essays and Hymns of Synesius of Cyrene*. Vol. 2. Translated by Augustine Fitzgerald London. Oxford: Oxford University Press, [405 AD] 1930.

Saint Augustine. *Confessions*. Translated by Henry Chadwick. Oxford: Oxford University Press, [397–400AD] 1998.

"Subliminal Projection." *Advertising Age*. September 16, 1957: 127.

Sully, James. "The dream as a revelation." *Fortnightly Review* no. 59 (1893): 354–365.

Thigpen, Corbett H., and Hervey M. Cleckley, "On the Incidence of Multiple Personality Disorder: A Brief Communication." *International Journal of Clinical and Experimental Hypnosis* 32, no. 2 (1984): 63–66.

Thorndike, Edward L. *The Principles of Teaching: Based on Psychology*. New York: Mason Press, 1906.

Thurber, James. *My World—and Welcome to It*. New York: Harcourt, Brace and Company, [1939] 1942.

Tulving, Endel, and Daniel L. Schacter. "Priming and Human Memory Systems." *Science* 247, no. 4940 (February 1990): 301–306.

Tylor, Edward B. *Primitive Culture: Research into the Development of Mythology, Philosophy, Religion, Art and Custom, Vol. 1*. London: John Murray, 1871.

Vivekananda, Swami. *Raja Yoga*. Leeds: Celephaïs Press, [1896] 2003.

Wallas, Graham. *The Art of Thought*. London: Jonathan Cape, 1926.

Walter, William Grey. *The Living Brain*. Harmondsworth: Penguin, 1953.

Watson, John B. "Psychology as the Behaviorist Views it." *Psychological Review*, 20 (1913).

Watts, Alan. *The Book: On the Taboo Against Knowing Who You Are*. London: Abacus, [1966] 1973.

Wayland, Francis. *Elements of Intellectual Philosophy*. New York: Sheldon and Company, [1854] 1869.

West, Louis J., et al. "The Psychosis of Sleep Deprivation." *Annals of the New York Academy of Sciences* 96, no. 5 (1962): 68–69.

Wigan, Arthur Ladbroke. *The Duality of the Mind: A New View of Insanity*. London: Longman, Brown, Green and Longmans, 1844.

Winnicott, Donald W. *Playing and Reality*. London: Routledge, [1971] 1991.

Wodehouse, P. G. *Right Ho, Jeeves*. Harmondsworth: Penguin, [1934] 1978.

Wood, Wendy. *Good Habits, Bad Habits*. London: Macmillan, 2019.

INDEX

Publisher contact:
The MIT Press
Massachusetts Institute of Technology
77 Massachusetts Avenue, Cambridge, MA 02139
mitpress.mit.edu

EU Authorised Representative:
Easy Access System Europe, Mustamäe tee 50,
10621 Tallinn, Estonia
gpsr.requests@easproject.com

Printed by Integrated Books International,
United States of America